"As a retired naval officer, I found *Wisdom from the Wild* the most engaging book on 'change'—both from an organizational perspective and as an individual leader—that I have read in years. Julie's correlations of change between something physical and something logistical are spot on. An easy read with concrete steps on how best to manage change within any organization, I highly recommend this book!"

—GREGG HUDAK

"In Julie Henry's book, *Wisdom from the Wild: The Nine Unbreakable Laws of Leadership from the Animal Kingdom,* come key lessons in leadership for any type of organization. These lessons are creatively gleaned from nature's animals ranging from the innocuous naked mole rat to the elegant cheetah to the gutsy sea cucumber. With each animal, Julie showcases what leaders can glean from these vastly different organisms in terms of their leadership style, improving their leadership qualities, and informing their leadership practices. Through the lenses of change, teamwork, and resilience, Julie does a superb job of describing why these lenses are vital to a healthy ecosystem in leadership. This book is not your typical leadership book, but one that provides powerful metaphors to examine our leadership practices. It is a timely, creative book that is both personal and applicable to all organizations, ranging from nonprofit to for-profit. This book will linger with you long after you complete it and reflect on its lessons."

—ANN HALEY MACKENZIE, ASSOCIATE PROFESSOR, MIAMI UNIVERSITY (OH), AND FULBRIGHT SPECIALIST

"A career of exploring the animal world has given Julie a unique and introspective perspective of the leadership lessons we can learn from wildlife. The instinctual nature of animals she describes pairs brilliantly with our drive to succeed and lead in our own way. I love how this book reminds us that our leadership capacity is a part of our inherent animal nature."

—BILL STREET, SENIOR VICE PRESIDENT OF INDIANAPOLIS ZOO AND DIRECTOR OF THE GLOBAL CENTER FOR SPECIES SURVIVAL

"*Wisdom from the Wild* is a true page-turner for those wanting to hone their leadership skills. Julie does a great job of simplifying the truly important aspects of leadership and how to make them a reality in your professional setting. I highly recommend this book as a relevant and 'go-to' book for anyone in a leadership role."

—HEATHER KASTEN, PRESIDENT/CEO FOR THE GREATER SARASOTA CHAMBER OF COMMERCE

"A lifelong passion for animals, nature, and our zoological backgrounds connected the author and me decades ago. I can attest to at least some of Julie's animal experiences—who wouldn't want to hold a tarantula? So how could I not be curious about *Wisdom from the Wild*? What I found are fascinating and unexpected connections between Julie's chosen species and leadership lessons, carefully crafted and communicated through stories from Julie's own life experiences blended with her expertise. Recognizing traits in some unusual plants and animals that make for a thriving planet—and applicable to us as individuals, team members, and leaders—is ingenious—and helpful—to anyone with a thirst to learn more about nature and themselves."

—JULIE SCARDINA, PRESIDENT/CEO, MISSION WILDLIFE

"*Wisdom from the Wild: The Nine Unbreakable Laws of Leadership from the Animal Kingdom* is destined to take its rightful place among other essential leadership references such as *Good to Great*, *Forces for Good*, and *Start with Why*. This book presents the fundamentals of leadership using unexpected role models from the natural world—both plants and animals—as sage guides, experienced mentors, and inspirational teachers, encouraging us to be the leaders we are meant to be. Henry tackles universal leadership challenges—leading through change, leading teams, and resilient leadership—offering guidance from unlikely but evolutionarily tested sources, from the steady, sturdy coastal mangrove forest to the tenacious and improbably loveable naked mole rat. This is not your ordinary leadership how-to. Henry's approach is a masterful combination of familiar leadership challenges presented alongside surprising examples from Mother Nature herself. Ranging from the seemingly mundane to the undeniably mysterious, Henry offers unforgettable examples from the natural world, meant to spur meaningful shifts in attitude and behavior to take your performance and impact as a leader to the next level. Whether your work as a leader is centered in nature or far from it, this book is written for you."

—RACHEL BERGREN, EXECUTIVE DIRECTOR, GET OUTDOORS NEVADA

"There has never been a better time to follow Julie on a journey through the wild to change the way you view the world. Julie's advice and insights show leaders how to re-connect with nature to inspire employees and drive innovation in the workplace. Julie's passion for unusual creatures and vibrant outdoor spaces is embedded in her DNA, and these leadership lessons are an absolute gift to be treasured and shared as you hike through life's challenges and opportunities."

—CATHY CHAMBERS, SENIOR CORPORATE LEADER

"This story of a woman's personal journey, from shy teenager to confident nature lover, educator, and business trainer, draws metaphors from the natural world in a way that is both entertaining and enlightening. Her love of animals and the examples they provide for the human experience shine through in this engaging tutorial in leadership training. Julie Henry bares her soul and shows how all our challenges, as humans living in an artificial world of our own making, can be met if we just draw from the beautifully evolved world of Mother Nature."

—BOB HUETER, PHD, MOTE MARINE LABORATORY & OCEARCH

"This love letter to nature offers compelling lessons about how to lead through change through the most charismatic of teachers. With all of us trying to make sense of the world that is moving so fast, with change as a constant, it feels like a gift to consider what we can learn from a naked mole rat or a mangrove. We hear we should turn to nature to rest and recharge. Julie Henry shows we can also turn to it to inspire and teach. Thank you, Julie, for sharing the wisdom nature has to offer all leaders on our journey."

—KRISTEN GRIMM, PRESIDENT, SPITFIRE STRATEGIES

"At a time when protecting the planet requires bold leadership, Julie brings us to nature to find inspiration and guidance to navigate the challenges of the leadership landscape. Through powerful insights and uniquely personal stories from her experiences as a leader in the conservation field, Julie takes readers on a journey to connect with the wild world while exploring critical lessons to adapt to change, maximize impact through teamwork, and build resilience. Leaders of all types will leave with a roadmap for growth and a desire to get back out into nature!"

—SEAN RUSSELL, FOUNDER AND DIRECTOR, YOUTH OCEAN CONSERVATION SUMMIT

"*Wisdom from the Wild: The Nine Unbreakable Laws of Leadership from the Animal Kingdom* is unique. It effortlessly weaves personal and animal stories together into leadership strategies and lessons. I found myself following stories about turtles, spiders, and a marathon, wondering how it would relate back to leadership and then receiving a satisfying 'aha moment' at the end. Ms. Henry also uses questions at the end of the sections to ensure the reader uses the lessons learned for their own reflection and personal accountability. I highly recommend this book."

—CHARLISA A SHELLY, SHRM CP, CHIEF PEOPLE OFFICER, KANSAS CITY ZOO

"Who would have thought that mangroves and giant spiders could teach us about leadership? Author Julie Henry takes a creative and personal journey into leadership using the natural world as her guide. From managing change to building teams, she shares new insights about leadership drawn from the natural world, which always has something to teach us."

—JUDY BRAUS, EXECUTIVE DIRECTOR, NORTH AMERICAN ASSOCIATION
FOR ENVIRONMENTAL EDUCATION (NAAEE)

Lily—
Keep leading in
the way you
were born to
lead.

WISDOM

FROM THE

WILD

Enjoy :)

JULIE C. HENRY

WISDOM

FROM THE

WILD

The NINE UNBREAKABLE LAWS *of* LEADERSHIP

from the ANIMAL KINGDOM

GREENLEAF
BOOK GROUP PRESS

This publication is designed to provide accurate and authoritative information in regard to the subject matter covered. It is sold with the understanding that the publisher and author are not engaged in rendering legal, accounting, or other professional services. Nothing herein shall create an attorney-client relationship, and nothing herein shall constitute legal advice or a solicitation to offer legal advice. If legal advice or other expert assistance is required, the services of a competent professional should be sought.

Published by Greenleaf Book Group Press
Austin, Texas
www.gbgpress.com

Distributed by Greenleaf Book Group

For ordering information or special discounts for bulk purchases, please contact Greenleaf Book Group at PO Box 91869, Austin, TX 78709, 512.891.6100.

Design and composition by Greenleaf Book Group and Brian Phillips
Cover design by Greenleaf Book Group and Brian Phillips
Cover images copyright Francesco_Ricciardi. Used under license from Shutterstock.com, ©iStockphoto.com/Jasmina007, and ©iStockphoto.com/Alkalyne
Original chapter illustrations by Kevin Stone

Publisher's Cataloging-in-Publication data is available.

Print ISBN: 978-1-62634-886-8

eBook ISBN: 978-1-62634-887-5

Part of the Tree Neutral® program, which offsets the number of trees consumed in the production and printing of this book by taking proactive steps, such as planting trees in direct proportion to the number of trees used: www.treeneutral.com

TreeNeutral

Printed in the United States of America on acid-free paper

21 22 23 24 25 26   10 9 8 7 6 5 4 3 2 1

First Edition

"We delight in the beauty of the butterfly, but rarely admit
the changes it has gone through to achieve beauty."

—MAYA ANGELOU, POET, AUTHOR, CIVIL RIGHTS ACTIVIST

. . .

"Just because you are CEO, don't think you have landed.
You must continually increase your learning, the way you think, and
the way you approach the organization. I've never forgotten that."

—INDRA NOOYI, BOARD OF DIRECTORS, AMAZON;
FORMER CHAIRMAN AND CEO, PEPSICO

THE TABLES TURNED

by William Wordsworth

Up! up! my Friend, and quit your books;
Or surely you'll grow double:
Up! up! my Friend, and clear your looks;
Why all this toil and trouble?

The sun above the mountain's head,
A freshening lustre mellow
Through all the long green fields has spread,
His first sweet evening yellow.

Books! 'tis a dull and endless strife:
Come, hear the woodland linnet,
How sweet his music! on my life,
There's more of wisdom in it.

And hark! how blithe the throstle sings!
He, too, is no mean preacher:
Come forth into the light of things,
Let Nature be your teacher.

CONTENTS

FOREWORD

I have known Julie for over twenty-five years and had the pleasure of her working for me when I was the CEO/president of the John G. Shedd Aquarium in Chicago. Over my forty-year career in the zoo and aquarium world, and amongst the thousands of employees I have managed, Julie stands out due to her enthusiasm, dedication, knowledge, and love for animals, large and small.

What makes this book such a valuable educational tool as well as an entertaining read? It's the creative way Julie has woven the theme of resilience, change, and teamwork through nine examples in the animal world along with her personal experiences and interests. This book is an opportunity to convey a message that matches almost perfectly with what different associations are engaged in while training future leaders. Certainly we can learn from *Wisdom From The Wild*.

Julie has made the biggest impact through her work incorporating insights from professional visits to over sixty organizations such as zoos, aquariums, nature centers, botanical gardens, and other environmental frontiers. She has participated at national and international conservation conventions and provided counseling services to numerous zoos and aquariums, along with corporate clients in manufacturing, hospitality, tourism, and insurance, as well as government and other nonprofits.

Julie has two children, and on any day, I can imagine they are ready for the latest "Mom Adventure." It's off to the mountains, the lakes, rivers, or even to the oceans and coral reefs. Her children have taken on her love for the wonderful outdoors and the fascinating animal world. The leadership she provides to these experiences will last a lifetime.

I am honored to write this foreword. Enjoy the book!

—TED A. BEATTIE, RETIRED CEO/PRESIDENT,
JOHN G. SHEDD AQUARIUM, CHICAGO, ILLINOIS

PREFACE

This book has been twenty-five years in the making. Clasped in your hands is the culmination of an idea that started to germinate during my senior year in college and, over the next two and a half decades, would be refined through my experiences gained as a leader of people . . . case studies learned as a consultant working across industry categories . . . global perspectives assimilated through time spent living and traveling abroad . . . and successes, failures, growth opportunities, and learning experiences too numerous to count.

In short, this book was first a skeleton of an idea.

Now, it is an idea with meat on its bones.

This idea lives and breathes with time-tested applications and process-driven scars. It has grown stronger as I have learned what does *not* work as readily as I have learned what does. It's an idea that has grown and evolved over time with nuggets of wisdom that can be earned only through boots-on-the-ground experiences.

It's a love letter of sorts to my original, beloved profession—one that expresses how much I adore working within, alongside, and rooting for zoos, aquariums, nature centers, wildlife rehabilitation centers, and conservation organizations of all shapes and sizes.

It's a gathering of the stories I now tell to my clients, from all types of industries, who may have never worked anywhere near wildlife and wild places, but who can share in the magic of the learning from nature nonetheless.

It's a collection of the experiences, the memories, and the moments that I hope will challenge, inspire, and motivate you to continue growing as leaders.

It's a tribute to all the people I've worked for, collaborated with, and led, who have shaped me into the professional I am today.

Even the process of actually writing the words in this book was essential in generating and fine-tuning the ideas contained in the work you hold in your hands today. Although I've been thinking, learning, applying, refining, and evolving these concepts for a very long time, the process of putting pen to paper provided the crucial distance that allowed me to intentionally decide how best to share with you the lessons I have learned from animals over the years.

On a quest to ensure equal representation in this book, I made purposeful decisions to include animals from both land and aquatic ecosystems. Working diligently to uncover the lessons of wildlife from a variety of taxonomic categories—we'll learn from mammals, birds, reptiles, fish, invertebrates, and even plants and algae—I cast my net wide, to include relatively unknown creatures alongside the known. (Sorry, amphibians! You'll see yourselves represented in the next book.)

I chose from my experiences working inside and visiting more than sixty zoos and aquariums, but also from nature centers, conservation organizations, and wildlife rehabilitation centers around the world. I picked stories I've been a part of while working as a consultant with people in industries ranging from public administration and insurance to manufacturing and tourism. I've also included tales from my personal time spent exploring the outdoors, whether it was hiking in New Zealand or snorkeling in the Florida Keys.

Nature is always teaching us, if we have the presence of mind to listen.

I learned early in my career that deep down inside, people understand this—which is why they always want to hear "one more story" about wildlife and wild places and the lessons they can teach us as leaders.

Nature has taught me to be a leader in ways that no textbook or mentor ever could. Today, I fully recognize my growth as a professional and as a person over the years I've spent learning from animals. I've been terrified to fall backwards over the side of a boat at night, off the coast of the Big Island of Hawaii, wearing heavy scuba equipment—but purposely fall I did, too excited to see the manta rays swimming below to let fear stop me. I've jumped with fright at unfamiliar sounds while hiking in Australia—only to find myself overjoyed at discovering that those heavy sounds belonged to a group of kangaroos hopping by.

And while the lessons I will concentrate on in this book center around the concepts of change, teamwork, and resilience—traits that are so readily prevalent in nature as well as in the life of an organization—the earliest leadership lesson I remember learning from nature was about embracing a choice that must accompany these traits: how to be brave.

If you were to meet me today, you would most likely describe me as outgoing and highly energetic. If I were delivering a keynote speech or facilitating a strategic visioning workshop, you might assume that I'm the type of person who craves being around other people and leads a life filled with group activities and large circles of friends. Although I have great respect and admiration for this type of person, and I certainly love and need the people in my life, the side of me you would be observing is a set of learned skills. In reality, my natural inclination is best captured in my preschool photo—in which I gaze at the camera shyly, without showing any hint of a smile.

I was the child who, during the kindergarten play, hid offstage when it was my turn to go on. It was impossible for the audience *not* to see me, clad in an oversized paper flower headdress, yet there I cowered—frozen in place, unable to walk out in front of people or even open my mouth.

I was the kid who, when it was my turn to race for the middle school swim team, climbed up onto the swim blocks and then promptly climbed back down before the starter gun went off, leaving my coach searching for me and wondering who would swim in my place.

I was the high school student who, harboring a deep love of singing and adoring her choir teacher, gathered up the courage to audition for the elite choir—but when asked whether I could read music, I inexplicably turned red in the face and answered, "No." At home hours later, I cried to my mother, devastated—because I had been able to read music for the past ten years.

I was the science-loving college student who chose to study zoology but was not interested in being a doctor and not all that good at being a research scientist, leaving me with no idea what I would actually do for a career.

I felt lost.

Until one day in 1995, while visiting a friend, I happened to pick up a *Discover* magazine lying on a table, with a cover feature titled "Playing God at the Zoo."[1] Interest piqued, I flipped to the designated page and started reading about all the challenges and decisions that zoo leaders have to make on a daily basis for the well-being of their animals and future of the species.

I was an awe.

The next two years until my college graduation flew by. With my newfound passion for zoos, aquariums, nature centers, and conservation organizations, I sought out every possible volunteer experience and applied for a number of internships. I worked with birds at a local rescue center, learned how to handle snakes for a YMCA camp, earned my scuba credentials to take students diving on a coral reef, and counted fish eggs under a microscope in the middle of the night to study their life cycle.

Everything was rolling forward, and I was confident my future after graduation would involve working for wildlife and wild places. I was especially keen to start my career with a job at my favorite aquarium, world-renowned for its conservation work and visitor experiences. I was so excited; I could hardly stand it.

The moment arrived at last, and the timing could not have been more perfect: I was due to graduate in three weeks. I applied to the aquarium and, upon receiving an offer to interview, worked hard to prepare for the big day. After the interview, I met various people on staff and then traveled back to my apartment, feeling nervous, excited, and confident that the first phase of my professional life was beginning to emerge. The interview had gone exceptionally well, and the anticipation was unlike anything I had experienced before—thinking about it now still makes my hands sweat.

This was long before cell phones and text messages, so I was bound to my landline. The hours clicked by, and then the days. I held my breath every time the phone rang, certain the call was for me. Finally, I arrived home one day and saw the blinking light on the answering machine. This had to be it! I took a breath and pushed the button.

"Hi, Julie, I'm calling from the aquarium," came the woman's pleasant voice. "It was lovely to meet you and we really appreciate you coming up to interview. You have exceptional credentials and will go far in your career. But unfortunately, you are not the right fit for us at this time. We wish you the best of luck and happy holidays!"

I sat in stunned silence.

All the little girl moments in me started to rise up—all the times when I had not taken a chance, when I had been afraid, when I couldn't speak up in front of people. This was the little girl I had been all those years. And suddenly I was not willing to be that little girl anymore.

This was my *dream job*. This was my chance to work at something bigger than myself, something that could have a global impact for wildlife and wild places. This was a career I believed I would deeply love and a purpose that I could sink my teeth into for the long haul. This was an organization that I had followed my whole life and one I desperately wanted to work for. I was not ready to walk away from my dream job. I was not ready to accept that I was not the right fit. I (*gasp!*) thought they were wrong.

I don't know what arose in me. But I do remember distinctly telling myself that I was not ready to give up yet.

So, I picked up the phone, called the aquarium, and asked to speak to the person I had interviewed with. I asked why she did not think I was a good fit and listened to her reply. Then, somehow, I found an answer.

"I hear what you are saying," I said. "But I feel our interview went well. I believe I could bring deep value not only to the aquarium, but also to your team. I am a hard worker, and I learn fast. I think I could be a good fit if you give me a chance."

Truthfully, I didn't even recognize my voice or the words coming out of my mouth, as this was so uncharacteristic of how I had lived my life up until that point. But there was something about *this* job, *this* aquarium, *this* opportunity, *this* moment, that I was not willing to walk away from so easily. I had picked up the phone to fight for myself.

"Tell you what," I continued. "I know you haven't called my references. Could you please at least call them and ask them their opinion? And if they don't agree, or if you don't like what they say, then OK. I'll agree that I'm not the right fit. But if that's not the case, then I think I will make an excellent addition to your team."

Now both she and I were stunned into silence. A wave of courage had swept over me as I thought about all the possibilities in store for me and what this opportunity could mean for my future. I had a choice to make. I chose to be brave.

The aquarium manager agreed to do what I had asked and contacted my references. Two days later, I got the call—and the job. And a few short weeks after that, I was on my way.

These are the sorts of moments that make us who we are, the moments when we choose to lead ourselves first before we earn the right to lead others—the moments when we dig in and change the trajectory of our path, when we choose bravery over blanket acceptance of what is, and wonder instead about what could be.

Nature has taught me how to do this—how to keep going along a path when I cannot see the end. Nature also has been my solace when things have not worked out the way I wanted, even after harnessing bravery. And nature has nudged me to choose bravery again the next time, to face down my fear of the unknown and of that which I cannot control, to choose opportunity and transformation.

I have spent the past twenty-five years writing and rewriting sections of this book in my head, refining these ideas at any chance I could: while leading my teams, helping my clients, delivering keynote addresses, facilitating workshops, and doing a thousand other things. And I have spent these years choosing to walk outside, take a deep breath, and dive beneath the surface of the water, all to discover new ways of doing things, new perspectives on old problems, and new sources of comfort and rejuvenation. At every step along my journey, I've worked to purposefully tap into my individualized approach to familiar challenges, motivated by observing how wildlife does the same thing. In watching an animal's instinctual behavior in action, I've been encouraged to listen to my own instinctual leader within.

We can all choose to listen to our internal voice, which guides us instinctually in ways no other person ever could.

We can all choose to leave our comfort zones and sit outside, observing and learning what only nature can teach us.

I hope you are ready to lean in a little closer and trust this voice of yours.

I think you are ready to take a walk on the wild side.

I believe you are ready to unleash the instinctual leader within.

AUTHOR'S NOTE

I started writing this book in the fall of 2019 with a plan, expecting it to take a certain amount of time and scheduling it accordingly into my work-flow. I was excited to finally put all of my ideas, case studies, and animal stories down on paper. I was excited beyond words to share nature's lessons about leadership and how we can look at change, teamwork, and resilience through a new lens.

Then the coronavirus hit, and the world—and life as we knew it—turned upside down.

I was in Southern California working with a client one morning at the end of January 2020. He explained that he looked so tired because his wife was dealing with the flu. And not only was *she* sick, she was extremely worried that *he* was also going to get sick, with some "new" type of flu: the coronavirus, whatever that was. As it turned out, my client had been in Wuhan, China, a few weeks earlier.

We laughed it off and went about our day as planned, helping his team create a strategic plan for the future of their department. We identified what made their work and impact unique while simultaneously isolating processes and decisions that could support the team in becoming more efficient. Ultimately, our goal was to help him drive lasting change.

Within a few short weeks, the pandemic was declared and lockdowns kicked in. Suddenly, the idea of having the "flu" took on a whole new meaning and my face-to-face time in California seemed a lifetime ago. My client and I began connecting virtually in what would become the norm rather than the exception. Thankfully, his wife recovered, and he never got sick (neither did I!). But our off-the-cuff laughter had been replaced with a sobering reality of how completely—and how suddenly—the world had changed.

Over a year has passed, and I still haven't been on an airplane since that trip to California. As is the case with millions of people around the globe, everything has changed for me. I am now a business owner/online home-school teacher/mask wearer/stay-at-home-longer-than-ever mom/Zoom happy hour enthusiast. Yet what has hit home for me most of all is the idea that change is no longer a luxury.

The idea of "change management" as a concept, or even a discipline, seems ludicrous and indulgent. Don't misunderstand me: I firmly believe in and embrace the goals of planning and coordinating change throughout companies of any size. I am certain there will always be a need for terms, deliverables, and coordinated efforts toward achieving those goals. But beginning in early 2020, the change that started circling and impacting us all on a daily basis didn't afford us that time. Change began impacting every single facet of our lives, from what we did for a living to how we actually lived each day, almost instantaneously. What began as a reaction to disruptive change, on a scale none of us had experienced, moved out of necessity toward a proactive focus on adaptation. Suddenly, the ability to change was required so we could not just survive, but truly thrive.

And this leads us right back to the beginning.

Now more than ever, I believe we need to learn the lessons that only nature can teach us. Just this week, I had a conversation with a colleague during which he described the coronavirus as a forest fire that is effectively "killing" our old way of doing things and forcing us to evolve. The aftermath of a forest fire offers opportunity for new growth in ways that were not probable or even possible before the fire. And my colleague is right: I can't think of any industry, in any corner of the world, that will be able to go back to business as usual. We all must creatively adapt to our new reality and embrace the growth opportunity made possible through the tragedy of the fire.

Leaders have never been so important. The ability to make decisions in a time of unprecedented unpredictability is a treasured and sought-after trait. No one alive has gone before us on this path—the pandemic has leveled the leadership playing field. Now more than ever, every decision has to be made using a combination of facts, observation, and input, in addition to trusting your gut and standing behind those decisions as the results play out in real time.

On account of the pandemic, we have also been deeply reminded of our intrinsic link to nature—that we are truly all connected—and of the need to not just celebrate but respect this connection. This perspective is well understood among conservation and nature-minded professionals, but the coronavirus has thrust it into the mainstream. Business and nature are interwoven in ways we cannot ignore and—as we've learned the hard way—if we don't focus on both, there will be a price to pay.

Even as we all become adjusted to a "new normal" in our lives, nature continually shows up in both expected and unexpected ways. We may be equally steeped in fear of the unknown and in gratitude for what we once took for granted in our busy lives: our health and the people who work every day to keep us healthy. One organization demonstrated this notion of coexistence and the impact of nature beautifully and meaningfully, when Barcelona's Liceu opera house gave its first pandemic-era concert: to an audience of 2,292 plants, livestreamed for viewers around the world to enjoy. Afterward, the plants were donated to 2,292 hospital professionals working daily on the front lines during the pandemic.[1]

It is my hope that this book, too, holds great beauty and meaning—now more than ever—as you lead your company, your community, or even your family. I hope it gives you new ideas and inspiration, and maybe creative ways to think about challenging problems. Most important, in this time of great uncertainty, I hope this book brings you the comfort and reassurance of knowing that you—like the animals—are wired not just to survive, but to thrive in the face of unprecedented change.

INTRODUCTION

Leadership: At its core, leadership is about building, earning, and keeping trust.

At least, that's what the leadership experts say—those who have studied and worked for decades with leaders of companies, communities, and countries. And while I believe that fundamental statement is true, both from my personal experience of being led and from being a leader myself, I believe there is a second fundamental truth at the heart of leadership.

Leadership: At *your* core, it is the instinctual knowledge that can guide your decisions and help you lead the way *you* were designed to lead.

There is a time and place for sitting back and learning from the experts, for reading the academic papers and case studies, which will inform the direction you choose to take. There is a time and place for seeking input from your team and listening to your mentors, for carefully weighing options presented by colleagues and considering the decisions of leaders who have gone before you.

And then there is the time to move forward. To make the best decision *you* can make with the information at hand, and then stand by the outcome. To bravely take a step back and let others make a decision they may be more qualified to make. To support the direction chosen by your leaders and implement the change you see in front of you, and to choose to trust the process as it unfolds.

In truth, leadership is an ever-fluid balance between these two scenarios—input gathering and decision-making. It is a sliding scale that shifts back and forth with each situation, each client, each financial impact, and each stakeholder input. Yet regardless of the process at hand, the decision to make, or the eventual outcome, a decision *must* be made and progress *must* continue to happen. Things—both people and processes—must move forward and, by definition, will change over time, as each decision is made. This is the scientific concept of evolution.

That's what makes nature such a perfect teacher for the challenging and very personal concept of leadership. Because each animal (even in the same species!) may approach a problem differently, the associated outcomes will vary accordingly. And because animals are probably not overthinking the way humans may be prone to do, animals make an instinctual judgment call and live with the outcomes.

So, here's the point: *You* are the one making the leadership decision. The well-published author or your mentor from college is not standing over your shoulder, and even if they were, they wouldn't know the situation as well as you do. It's time to trust your instinctual nature.

To help you do that fully, I want to share with you how to learn from and be inspired by the wildlife and wild places all around you. Wise insights and novel approaches to leadership challenges are just waiting to be discovered—along with a healthy dose of motivation and restoration to help you lead from the inside out.

If you are an experienced leader who has been at the top of your game for a while and you are looking for a creative way to explain leadership to your successor or team, I've written this book for you—so you can approach conversations and possibilities with innovative examples.

If you are a productive leader who is feeling overworked, is stretched thin, or needs a jolt of energy, I've written this book for you—so you can fully appreciate that no one can function at peak performance forever, and so you can take the much-needed time to rest, recharge, and come back fully ready to move forward.

If you are a new leader at any stage in your life or professional journey, I've written this book for you—to help you balance the potential onslaught of leadership advice with listening to your own instinctual nature, in order to bring to life your own ideas about growing your team or expanding your impact.

And if you are a person who loves animals and nature as much as I do, I've written this book for you. Some of you may be working in your dream field, which includes interacting with animals and nature on a daily basis. But far more likely, based on the people I've met throughout my career, you're working in a different profession yet still want to maintain a connection to nature—whether that's through seeing wolves in the wild or simply walking your dog on the weekends. I hope this book can be a way for you to incorporate your inherent love and respect for the natural world seamlessly into your job. (And it may also provide an excuse to take your team on a retreat to the local nature center or aquarium—or even a way to share more stories about your cat!)

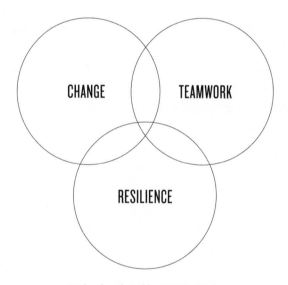

Wisdom from the Wild Leadership Circles

I've broken this book into three parts: **Change**, **Teamwork**, and **Resilience**. At every stage in your career, you will be leading *change*—on a big or small scale. The incredible honor of leading people means you need to continually learn and reflect on how to guide them as they work together in *teams*. And to lead at your highest level, you must build and nurture the *resilience* muscle in both yourself and those you lead. These three parts are basic fundamentals we can learn from nature that are immediately applicable to the business world. I also believe most (if not all) business lessons are applicable to every facet of life, which means you may be able to apply this wisdom from the wild to how your family interacts just as readily as to how your team interacts.

All living things—whether people or plants or snakes or squid—have basic needs: food, water, and habitat. Depending on the creature you're thinking about, additional things are important, such as various nutrients, air to breathe, places to raise young, and so forth. But the basics are universal: To survive (and thrive) living things need something to eat and drink, and a safe place to live.

All leaders—whether of large corporate teams in companies or small, community grassroots volunteers—also have basic needs: to guide change, lead teams, and be resilient. Depending on the leader you are, additional things are important, such as managing conflict, communicating with purpose, navigating succession planning, and so forth. But the basics are universal: To lead at your highest level and have the impact only you were designed to have, you need to fully understand change, teamwork, and resilience.

The Leadership Circles visually represent not only the layout of this book, but also the overall approach to what we as leaders can learn from the animal kingdom. If one circle becomes too large and overshadows the others, or is perhaps missing entirely, the whole relationship will start to become unbalanced.

Take away CHANGE? Without a process to lead change, you will become bogged down in uncertainty, and the people around you won't know how to effectively participate—and therefore they will not trust the process or outcomes of change.

Take away TEAMWORK? Without considering who you need on the team and how teammates are working together, you won't be leading teams—you will be herding cats.

Take away RESILIENCE? Without valuing and implementing strategies for resilience, you will burn out well before you generate the impact you can have as a leader—and potentially take down others with you as they follow your (non-resilient) example.

Each of the book's three parts comprises three chapters that incorporate ideas and tools you can implement immediately, supported by case studies and research I have conducted over many years. Alongside these are both personal and professional stories that reflect relevant lessons I have learned from nature. Each chapter is summarized at the end with Pro Tips, and each part concludes with an Instinctual Leadership Field Guide so you can reflect in real time and actively apply the concepts to your role as a leader.

The leadership development tools I share with you are designed to focus your attention, so you can think critically about the ideas presented and live out of this "leadership tool bag" long after you finish reading this book. I have used these tools in my work over the past twenty-five years, inviting leaders to take what works for them and leave the rest, and I invite you to do the same.

Because I'm both a scientist and an educator by formal training, I approach the world through a lens of questions, seeking how best to construct my own understanding. So, I'll work with you the same way—asking you questions throughout the book so you can truly learn, apply the concepts in your world, and create your own understanding. This is the best advantage I can offer as you continue along your leadership journey—to encourage you to *question*, to *apply* tools and *customize* the techniques to best suit you. I have no desire to tell you facts and statistics and then simply leave you to figure out how to incorporate them into your own world. Instead, my objective is to provide an opportunity for you to co-create and take away the *meaning* from these concepts that applies directly to you.

The tools in this book are neither another "system" to be implemented nor "steps" to help you achieve certain goals. They are methods and techniques that you can absorb into your overall approach and that can be applied in a wide range of settings, such as within your team as you are planning fourth-quarter targets or for yourself as you are training to complete a marathon. I'll even share insights gained directly from my clients and participants over the years as I've asked about their specific pain points and worked with them to apply various strategies. Above all else, these tools and techniques are designed to be simple and straightforward on purpose, so they can be easily explained and replicated—but they are not always easy to execute.

If there is anything I have learned over the course of my life, it is the fact that "simple" does not always mean "easy." In fact, it's usually quite the opposite. There's a reason that a well-seasoned main dish or a great martini tastes better when made by someone with experience; such an item may need only a few ingredients and even fewer steps to put it together, but only a discerning eye can wield the subtle art of balance and flavor. And while, for example, training a dog or going bird-watching may seem like a simple prospect, I know from experience that these activities are definitely not always easy.

Most fun of all, I believe, are this book's direct links back to wildlife and wild places, with each chapter framed by an Unbreakable Law of leadership that we can learn from the animal kingdom. I call them "unbreakable laws" because they are true, fundamental guidelines that can steer our work and our decisions, with concrete examples from wildlife.

How did I decide to use the term "unbreakable laws" when considering how to share the wisdom from the wild with you?

On a beautiful day not long ago, I was walking through a world-famous botanical garden, admiring the wide variety of plants and the team's innovative approach in their award-winning children's exhibit. Distracted by the fun, interactive components that invited kids to push and pull at buttons and levers, I stopped to stare at an unusual plant with a bent stem and a haphazard assortment of leaves. I could tell just by looking at the plant that it had faced some challenges—maybe it had been too close to other plants or too far away from a window for a while, and obviously it had been growing in a couple of different directions as it sought out the sun. But grow it did, seemingly not deterred by the tiring efforts it had already employed. It continued reaching, searching, laser-focused on receiving more sun—one of its basic needs—*no matter what.*

Suddenly the term to describe the link between leadership and nature jumped out at me. These efforts were an unbreakable feature of this plant that always seemed to be true no matter what it encountered—an Unbreakable Law. Some things in nature are just always true. Likewise, some things about leadership are always true. Thus, the concept of Unbreakable Laws of leadership from the animal kingdom, and the layout of this book, was born.

And don't miss the behind-the-scenes peeks into some of my favorite animal facts, outdoor experiences, and moments working in and visiting aquariums and zoos around the world. These conversational nuggets throughout the book are designed to provide you a much-needed "brain break"—a chance to laugh, reawaken your inner child, reinvigorate your curiosity, and simply wonder about the world around you. It's as if we were sharing a cup of coffee or a glass of wine, and you asked me to relate one of my favorite stories about wildlife and wild places. I call these moments "Behind the Scenes with KiwiE" because they represent the brain breaks I always feel when I take a breath and head outdoors for a new perspective or visit my favorite animals at my local zoo.

And the nickname KiwiE? That represents the personal and professional journey I have been on over the years. I started life as "Julie" but when my younger sister started to talk, she could never quite say my name—so to her (and the rest of my family), I became "E." Decades later, I moved to New Zealand, where the locals are sometimes affectionately referred to as Kiwis (after their national bird). Fully intent on becoming an active part of my new community, I linked my two identities together and adopted the nickname KiwiE.

I've written this book to be informative and applicable, with a touch of fun and inspiration. We look to nature for everything else—designing buildings, inspiration for art, protecting ourselves from injury, medicines we have not yet dreamed up, even novel ideas for skincare. Why not leadership?

I hope this new approach to leadership challenges will help you solve problems and move your team forward. And as much as I love eagles, elephants, and lions, I believe it's relatively easy to look at these large, regal animals and impart lovely quotes and ideals that can apply to leadership. I'm far more interested in the animals that may be all too easily overlooked or may even be completely unknown to you. We'll focus our time on learning from creatures and plants that might seem unusual or even unexpected in a book about leadership, such as mangroves, naked mole rats, spiders, and sea cucumbers.

So, let's dive into the natural world as you have never seen it—full of lessons on how to deal with change, work more effectively as a team, and build your resilience muscle. I promise you'll never look at a giraffe—or a termite—the same way again.

Behind the Scenes with KiwiE

Octopuses are master escape artists.[1] The giant Pacific octopus can grow up to 16 feet in length and weigh 110 pounds (the largest individual on record was 30 feet in length and 600 pounds!),[2] but without bones, they can squeeze through an opening the size of the only hard part in their body—their beak. They've been known to move between exhibits at night in aquariums and enjoy eating a few fish from other tanks before returning back to their own habitat.

Aquarium volunteers once told me about the time they were having their morning meeting and getting ready to kick off their day, when an octopus walked into the room. Well, maybe not "walked," but deliberately moved its large, squishy, mucus-covered body across the floor.

Sure enough—around the corner in the aquarium? An empty exhibit where the octopus should be. Inside the small conference room? A few more elevated heart rates and wide eyes than a normal Tuesday morning.

CHANGE

"IT IS NOT NECESSARY TO CHANGE. SURVIVAL IS NOT MANDATORY."

—Dr. W. Edwards Deming, engineer, statistician,
Total Quality Management (TQM) expert

Change.

The one constant in life.

The one thing you can always count on.

The one thing that will still occur no matter how much you manage, lead, mitigate, prepare, avoid, or refuse to accept. Change can happen to you, because of you, with you, or in spite of you. You can greet it with a smile and open arms, or stare it down with angry eyes and a defensive stance. Nevertheless, the next corner you turn, you will still find it there—no matter what. Change is the way of life for all humans.

Guess what?

The same is true for animals.

Wildlife and wild places are not immune to change. In fact, they're designed to *adapt* to change—it's how they survive and thrive. The weather may shift abruptly, and the leaves may fall from the trees far earlier than they did the year before. Young animals can be separated from their mothers before they're ready and forced to defend themselves for the first time. A nest can be destroyed, and the inhabitants will need to build a new one before laying eggs. Or a new patch of land can be discovered by an animal family in need of a habitat.

The ability to adapt to or effect change is necessary at every stage in life. Sometimes this process is proactive, and sometimes it is reactionary. Animals can cause change as much as they can react to it. And the same is true for leaders.

From my clients over the years, I have heard everything from "change keeps me open-minded and flexible" to "change is continually happening—either adapt or die." They've told me that "change is about moving forward and growing; if you're not moving forward, you're falling behind," but if the process to develop and implement change is to be successful "there must be follow-up and accountability."

The following client quote captures the essence of how pervasive change is and why nature is the perfect teacher: "Change is CONSTANT! Whether at home or work, something is always changing. I'm always adapting and making adjustments as needed to get through the day or week or month. That said, I recognize the need to keep learning new ways to adapt in order to keep moving forward."

Leading, managing, and participating in change is sometimes described as a "soft skills" component of leadership. It's the people side of the job. Yes, there may be technical components that are necessary to the change, but it's really about human behavior. Harvard University, the Carnegie Foundation, and Stanford Research Center conducted a study that shows 85% of job success comes from having well-developed soft skills and people skills, while only 15% of job success comes from technical skills and knowledge (or hard skills).[1]

Let's be clear: Change always seems more palatable when it's your idea or you are able to participate in the process or decision in some way. When change happens to you or you feel completely cut out of the loop, it becomes that much harder to buy into or implement the idea. Will you always be the one leading the change? No, of course not. But you will become an even more effective leader of change (and be more able to make the change stick) when you remember the people you are leading have their own experiences with and visceral reactions to change too.

Anyone can adopt a proactive, focused response to change. Even if you seemingly have no control over it, even if it's not your idea or feels much larger than any challenge that's come before—you can choose to be in the moment. You can decide to stay present despite that uncomfortable feeling and choose forward progress over backwards retreat.

Leading change will take a quantitative approach to measure progress. It will require deliberate and intentional acknowledgment of, and communication about, ideas and decisions that may produce feelings of fear. You will need to commit (and then recommit) to forward motion every time, even when you can't see the end in sight.

Can you do this?

Yes.

Why do you need to?

Because "survival of the fittest" is not just for the animals anymore.

Behind the Scenes with KiwiE

Of all the things you could give me, please let it not be a plant. This might seem counterintuitive: a nature-loving person who cannot keep a plant alive. It's not that I haven't tried—I've attempted to grow all kinds of herbs and greenery, from finicky orchids to the hardy aloe plant. None have survived.

The last time I was gifted a plant, my kids rolled their eyes as if to say, *Not another one.* I, too, was unhappy at the prospect of killing yet another green, living thing. But to my great surprise and happiness, within the leaves I discovered a caterpillar! One day later, I found at least ten, and a few days after that, the plant was covered in them.

The plant, predictably, did not survive. But at least this time it was not my fault: The hoard of caterpillars took over and ate every last bit of plant matter they could. We found at least twenty-five chrysalises and soon observed as many butterflies emerge and fly away.

On the porch—one completely decimated plant. In the wild—dozens of happy butterflies. In our home—a happy family watching nature in action.

CHANGE IS CONSTANT, BUT IT DOESN'T HAVE TO BE CHAOS

(It All Comes Down to Roots)

Growing up in Chicago, I dreamed of the sea. I was extremely curious and wanted to explore the world to find out what animals lived where and how they perfectly adapted to living in different regions of the ocean.

As I started my career, I began having opportunities to fulfill my dream. I trekked with scientists over active lava fields in Hawaii to observe where lava poured into the sea, and learned about how this affected the animals and plants living nearby. I snorkeled through chilly water and giant kelp forests off the coast of California—maybe just a little terrified, but even more curious—searching for garibaldi, sharks, and sea otters that could be on the other side of every piece of kelp. Early one morning while kayaking with new friends off the coast of Dunedin, New Zealand, I almost dropped my paddle into the ocean when an unexpected animal swam up alongside my kayak: a penguin! To the surprise of my incredulous friends, for whom this was a common sighting, I quietly screamed in delight. Up until that point, it had been impossible for me to see penguins in the wild, as they inhabited an entirely different hemisphere from my home in the United States.

But the opportunity I never knew that I wanted came when I found myself snorkeling among mangroves along the Florida shoreline. I already had a healthy respect for these amazing trees, with massive roots that lend stability to the shoreline and provide critical nursery areas for animals—but that was all book knowledge, described in words on pages or viewed as static images. I didn't truly know what it was like underwater, among the mangroves, until I went there myself.

It was a bit creepy.

Easing over the side of the boat and into the dark water surrounding the mangrove roots was unlike entering the ocean near a coral reef, where I could typically see for hundreds of feet around me. Now I was in water with low light penetration and even less visibility. It was strange to be floating at the surface while at the same time still extremely close to the sandy bottom. Rather than recognizing big, familiar animals, I was now surrounded by tiny, somewhat hidden creatures that were much less familiar. It felt almost claustrophobic, so I had to tell myself to relax and take purposeful breaths. Once I did, I could start to truly see and appreciate what was around me.

Snorkeling in mangroves was part of my job, and I doubt I would have sought out the opportunity on my own. It certainly wouldn't have been my first pick if I were the boat captain that day. But once you discover a world that you didn't know before, it becomes captivating. You can't unsee the wonders you have seen from this new perspective.

Observing mangroves from under the water made them come alive for me in ways they never could have before. Their roots were much more than just haphazard brown structures that grow at seemingly odd angles from tree trunks. Working together as a dependable and cohesive system, they clearly formed both a nursery and an integral foundation for life in the sea, and stabilized the shoreline. The mangroves created a world all their own.

Observe Change from Underneath

Everyone has dealt with change.

Stop and think about the change you are dealing with right now. What comes to mind? What feelings does it bring up in you? Excitement? Anxiety? Is it under control? Spiraling out of control? Are you running toward the change with confidence or running from behind trying to keep up? Are you facing the change with confidence and gusto, or would you rather hide until it passes over? Are you surrounded by people as you approach this change, or are you facing it all on your own? Are you at the very beginning of the journey or nearing the end? Can you even tell the difference anymore?

Everyone has had these feelings or reactions at some point. At times, we may feel that the change is well under control—until something unexpected knocks us off course. At times, we may feel completely upside down in dealing with change, only to realize that we're actually right side up and reaching the conclusion.

Sometimes we have to react on the spot—change surprises us, and we must adjust. But sometimes we have days, weeks, months, even years to prepare for change that will come. We brainstorm and plan and delegate. Ideas are generated, budgets created, stakeholders interviewed, and media plans outlined. Speeches are written. People are mobilized. Operational excellence is targeted. You may catch a glimpse of this happening on the surface or have a vague idea of the scope, but the true breadth and depth of change occurs underneath the surface, where it's not always possible to see it or make sense of it.

But, at some point, all the thinking and planning turns into reality, and we are called to action. It is time to implement change. And if we're facing a complicated web of people and processes and ideas and plans, the organizational waters can seem clouded and chaotic. We can't always see how to move forward and make the necessary change; it becomes easy to feel overwhelmed and uninspired.

And when we're feeling stuck or lacking inspiration, what better place to look for it than in wildlife and wild places—specifically, in this case, in a world that may seem crazy and chaotic at the surface but has purpose among its roots: the mangroves.

Mangroves are by far my favorite trees on the planet—and not just because I can reliably identify them, but because of where they live (along the ocean's shoreline) and what they represent (stability and steadfastness). These trees have evolved in one of the planet's harshest environments for a tree—saltwater—to not just survive, but thrive.[1]

If you look at much of the coast of southern Florida, chances are you'll find mangrove trees anchoring the shoreline. These giant plants create entire ecosystems among their roots, a submerged network of branches and foliage that create a rich habitat and important nurseries for fish and other sea creatures. Jump in the water, and you'll discover a massive, hidden labyrinth of roots that filter the water and prevent the ocean's waves from eroding the land. From above, it may look like chaos; beneath, it's a system that's firmly rooted yet able to flex—it makes perfect sense.

Change may appear at first like this, unruly and unfocused. But look closer and you'll see ways to participate and opportunities to lead submerged, maybe even hidden, among the roots.

Perhaps, like me at times, you've watched from the sidelines, working under a leader while secretly craving some knowledge about why you're heading in a certain direction. I've also been the leader making decisions as fast as I could in pursuit of a necessary change, trying to keep the momentum headed in the right direction while pulling along my team.

As an outside consultant now, before I conduct any training or undertake a strategic project related to change, I always benchmark my work first by getting input directly from the participants—the people behind all the brainstorming, forecasting, scheduling, and preparing. I want to know all the submerged, hidden details: what they like about change and what derails them; what kind of change they're dealing with and how it's impacting their job; how they lead and/or manage change; and what they would like to do differently.

Here's what people universally tell me: They want to use a *process* that will help them both lead and participate in change. A process that creates a firmly rooted foundation, one that is durable in a variety of conditions and adaptable to wide-ranging needs. A process that we can learn directly from the mangroves themselves.

Root Your Efforts in a Solid Foundation

Three different mangrove species inhabit southern Florida's shoreline. The red mangrove is the first species you can see as it abuts the ocean, with giant "walking roots" that extend out into the sea. When seeds form, they start generating first on the tree (a form of "live birth") before dropping into the ocean to find a place to anchor, thus increasing their chances of survival. As the red mangrove starts to grow, the roots actively prevent saltwater from even entering the body of the tree. The leaves, too, work toward this goal; they feel waxy and thick to the touch, which helps prevent any water loss through the evaporation that might normally happen on tree leaves. The crashing waves can't penetrate this surface either.

Black mangroves grow right behind the red mangroves and have unique "snorkel roots" that pop up out of the ground, surrounding the tree. These roots allow the tree to take in oxygen from the air even as the tide comes in. The leaves of the black mangrove are distinctive—bright green on the top surface, but almost gray on the bottom. Running your finger over this surface betrays its secret: The tree allows saltwater to enter it, but then excretes the salt through its leaves. One lick, and you'll taste the sea!

The white mangroves are last in line, the farthest back from the sea but still growing in saltwater. Like the black mangrove, the white mangrove allows saltwater in but then actively excretes it through its leaves. When the thick, fleshy leaves fall off the tree, they take the excess salt with them. White mangroves are easily identified by their unique leaves, which resemble the classic Frankenstein's monster: the round "head" (leaf), the "neck" (stem), and the two "bolts" on either side of the neck (nectaries).[2] The nectaries were once thought to excrete salt, but scientists now know they excrete sugar—much to the delight of butterflies and other pollinators!

To me, the most fascinating thing about these three trees is how they work in conjunction to form a healthy, stable ecosystem along the shoreline. Their crucial importance to the habitat is recognized in laws that regulate how mangroves can be trimmed or removed by humans, but they have very little competition when it comes to nature. Whereas other species of trees and plants may be competing for places to live and elements for survival, mangroves occupy a niche that not many others are interested in. They have firmly rooted themselves in a foundation all their own. Without them, the shoreline destabilizes and the ocean loses a vital nursery habitat. If the mangroves disappear, the ocean can quickly erode any habitat formed.

As you embark on your journey to more effectively lead change, what is the foundation in which your efforts are rooted? How will you help people participate fully and effectively in the change you seek to lead? Do you have a quantifiable, replicable methodology in place by which you measure progress and growth—one that is easily shared with others and that helps ground your future path?

If not, you should.

Lead Change with the Mangrove Method

A trusted, time-tested methodology for leading change must begin with what is essential: easy to understand (and replicable) phases in which people can actively participate. This methodology is also cyclical in approach—which reflects the typical reality of change. Even if change is approached in well-ordered, consequential phases, the actual process of (and experience with) change is most often not linear. There are usually stopgaps, feedback loops, adjustments, and regrouping needed. A cycle can still direct forward movement—without the artificial (and potentially discouraging) constraints of anticipating a linear, one-way flow.

Borrowing from the three species of mangroves themselves, the Mangrove Method employs three distinct phases labeled A, B, and C.

Each phase is intended to be time bound and actionable with a chance for feedback and adjustment as needed. This is imperative for you as a leader, so you can implement time-bound deadlines and concrete measures of accountability, and the people participating in the change can trust a transparent methodology to know that the change is not going to go on indefinitely.

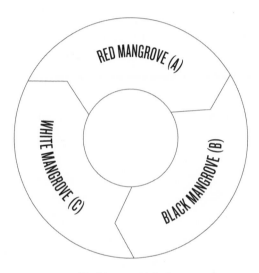

The Mangrove Method

The three species of mangroves found along Florida's shoreline do not grow in a straight line with definite separations between species. You'll find black mangrove roots popping up alongside an arching red mangrove root, or a white mangrove branch growing next to some grayish-black mangrove leaves. As you use the Mangrove Method, moving intentionally through the three distinct phases for leading (and participating in) change, there will inevitably be overlap: A new idea pops up, stakeholders want to participate, a law is enacted, or some market pressure demands attention. This methodology allows for flexibility in the process of leading change—just like the three mangroves adapt to the changing nature of the weather and the waves. But they remain firmly rooted, working together to anchor the shoreline, providing crucial habitat for the animals that live among them—just as you, even at times when the waves threaten to knock you over, will benefit from the foundational stability the Mangrove Method provides.

When the change you have the opportunity to lead is on the horizon, I want you to imagine that you are approaching the shoreline on a boat. You can see mangroves ahead of you. The first question of leading change is usually, where do I start? Well, which species of mangrove do you see first? You see the red mangroves, which are the first phase in our change methodology.

When change is coming, we all know we need a plan to address its challenges, so we start by, naturally, building the plan (Phase B). But by going right to building the plan, we've completely skipped something that we need to do first. It's not that we don't want to do this phase. It's just that, out of necessity or habit, we think it is best to get right to the perceived heart of the matter.

Imagine you are back in the boat with me, looking toward the shore. Black mangroves do not stand alone by themselves, absorbing the highest energy waves. The red mangroves are first. They are the ones extending their roots out into the relative uncertainty of the ocean.

If you don't know where you started, how will you know how far you've come? This is both a pragmatic and an emotional question; you need to start one step back to effectively measure progress, but also so you (and the team) don't dissolve into frustration. When time gets short and budgets need to be met, reflection on one's progress is often overlooked. So, let's start with the red mangroves first as you embark on leading change.

Phase A = Red mangrove

If you're familiar with mangroves, it will be easy to remember that this phase must come first. If you're new to mangroves, you can rely on the letters themselves to help you know what to do when. And in this case, Phase A is for *Assess*, the first phase in the Mangrove Method.

Assessment can take many forms, from an ultra-documented, formal process involving stakeholder input, written surveys, focus groups, and more to generating a list of minor improvements and new ideas that may require only one other person to buy into the process. If you are embarking on a strategic plan, rolling out a new company-wide system with official processes in place for change management, or undertaking a multiyear project, then your *assess* stage needs to mirror the depth and breadth of your efforts. But it's far more common for you to use this process while working on smaller tweaks and focusing on smaller goals, in which case the large-scale efforts to document the current state of affairs may not apply to you—but the need to assess still does.

The *assess* (red mangrove) stage of the Mangrove Method can take one of many forms: meeting minutes after a team discussion about what is going on in the department, a few notes taken during phone calls with colleagues around the country, or a search of internal documents to see what change was implemented in the past and where you stand today. Regardless of how official or unofficial your evaluation seems to be, here is the key: *Record and date your assessment.*

Life gets busy and memories get foggy. You will have to explain your progress to someone (your manager? an investor? a colleague?), and you need to know where you started to know how far you've come. Then before you move on, implement your built-in checkpoints using key performance indicators (KPIs) aligned with your goals and a feedback loop to ensure you've gathered and measured what you need to in this stage.

Now that the waves of change have been buffered by the strong roots of the red mangrove, we can move a little farther on shore. Here we'll find the black mangrove and the next phase in leading change.

Phase B = Black mangrove

You know where you came from, and where you hope this change will lead, so it's time to get back to doing—it's time to embark on Phase B and *Build* the plan.

Your plan could be a multi-page report that lives on the company website and is accessible to multiple departments. It could be a spreadsheet that outlines time-bound steps to be taken and who is responsible for each. It could be a one-page, process-flow document that helps capture the visual nature of where you are leading the team. Regardless of its breadth and depth, here is the key: *You must build a living and breathing plan.*

A document that sits on a shelf is of no use. A flowchart that can't flex to allow for changes in the market or client base is pointless. More times than not, some need or demand—some internal or external pressure—will come along to change your plan before you've even finished implementing it. As the leader, you must be savvy enough to recognize and roll with the need to evolve your plan as you measure your KPIs and listen to feedback, and you must ensure that your plan is flexible enough to stay relevant and current. This is why you started with an assessment to benchmark your work in the first place, so you can go back to the *assess* stage again and determine how to adjust your plan in order to continue making progress toward your goals.

The black mangrove will truly root your efforts in a solid foundation with a well-structured plan, but we're not yet finished. Now we find ourselves in the final stage, in and among the Frankenstein-looking leaves of the white mangrove.

Phase C = White mangrove

The final phase in the Mangrove Method is the letter *C*, which in this case is doing double duty (just like the two nectaries on either side of the white mangrove leaf!). After assessing the situation and building a plan, you must *Commit* to action and then *Communicate* the plan and the progress.

Why call out the idea of committing to action? Isn't it obvious, if you've gone through all the effort to assess and then build a plan, that you will follow through? I wish it *was* obvious. I wish I could tell you that all plans are considered, developed, and then executed, that none sit on the shelf collecting dust. But that would be a lie. I'd be shocked if you don't have at least one example of this yourself, whether of a plan you helped create or one that you attempted to lead. This happens even in our personal lives—just look at gym memberships and attendance. It's next to impossible to find a spot on the treadmill on January 4, but in just a few short weeks (or days), the plans created on New Year's Eve dissolve into the past.

Commitment and follow-through are the unsung heroes of leading change—the two aspects of a leader's skill set that are most often assumed to be a given, but that are also the most often overlooked or under-implemented. I believe leaders *want* to commit and follow through. Their intentions are in the right place. But sometimes, with so many plans in the air and so many changes coming down the pipeline, action becomes nearly impossible to take during the available hours in a day. So, the plan becomes what many people fear most: an exercise in futility.

What does commitment and follow-through look like? It starts with the leader leading by example: What parts of the plan do they need to implement first? It then occurs when deliverables are identified, people are tasked with specific responsibilities, and time-bound deadlines are agreed upon. It gathers momentum through weekly meetings, report-outs, accountability partners, and feedback routinely given to people in a timely matter. Sometimes this can be over a long time period (six months? three years?) and sometimes this is accomplished in a few weeks. Regardless, the commitment and follow-through are an equal part of the process we have been discussing—transparent, trustworthy, and time bound.

As this well-benchmarked and well-developed plan is officially on the way to implementation through commitment and follow-through, the last critical phase is communication.

Communication is not simply letting people know about the plan or giving them an update on progress. Truly purposeful communication requires proactively deciding *who needs to hear what message*. Each instance of communication is an opportunity to share with the audience what you need from them. Do they need to be aware of the plan and provide input, but not necessarily act? Or do you need them to change how they work or process information, so that new ideas can be implemented?

Purposeful communication also means considering *who should deliver the message*. You may be the leader—the person who is spearheading the plan and ultimately responsible for implementation—but that does not mean you are the best person to communicate the plan to every audience or stakeholder.

Regardless of your approach to the communication phase, here is the key: *You lead change by not just continually doing, but by committing to follow through and intentionally communicating with those around you.*

Assess (red mangrove), build (black mangrove), and commit and communicate (white mangrove)—these phases make up the Mangrove Method to leading change that sticks. Organizing these three discrete stages into a cycle allows you to act in a fully transparent manner and thus build trust in the process, in the outcomes, and in you as a leader.

Leadership Lessons in Real Life

Sometimes it takes a few tries before you actually learn something—but once you do, you can never unlearn it. Although this story happened over twenty years ago, I will always remember this moment as the first time I learned what leading change was really about.

As a young professional working at the aquarium right out of college, I would volunteer for every opportunity possible: teams to join, trainings to participate in, conferences to go to. I was hungry for more experience, looking to build my skills, and yearning to contribute in a meaningful way. I was ready to apply all of my book knowledge to a real, living-and-breathing organization full of interesting people, possibilities, and challenges.

A year into my tenure, the golden opportunity arose. A large coral reef habitat full of nurse sharks, sea turtles, and green moray eels was scheduled for renovation from the inside out, including developing new graphics and educational content, updating the interactive elements, and rewriting the daily dive show script. A highlight of any visit to the aquarium, this interactive show was hosted five times a day by a highly experienced volunteer who would scuba dive in the habitat wearing a positive-pressure scuba mask and talk to visitors in real time through a microphone. The diver would also feed the animals while explaining which creatures were which, sharing information about coral reefs and conservation, as hundreds of people circled the tank, faces pressed up to the glass. Visitors could even ask the diver questions via another staff member standing outside the tank with a microphone.

And I was the one chosen to lead the team in reimagining and rewriting the script for this signature dive show. I was ecstatic! Here was an opportunity to apply all of my learnings so far about leading change into one initiative, and I knew just what to do.

I started by building a team of both staff and volunteers, so we had representative stakeholders who could weigh in on any new ideas. We gathered data and assessed the status, content, objectives, and impact of the existing dive show. We employed creative brainstorming skills, even turning off the lights in one of the aquarium meeting rooms, lying on the floor, and imagining what it was really like to scuba dive a coral reef.

Then we built a plan for how we would create the next evolution of this show. We brainstormed, debated, pilot-tested, and reviewed multiple ideas to generate the best script possible. We received feedback from a variety of people on what would make the best show for the new habitat. Finally, after months of commitment and work, through countless steps, the script was ready to roll out to the staff and volunteers who would be responsible for narrating the show.

Ever the enthusiastic team leader, I couldn't wait until the night of the big reveal, a dinner presentation on all the changes and exciting new aspects of the exhibit and dive show. Halfway through the program, I took to the stage, extremely proud to represent my team and everything we had worked toward. But as I began explaining the reimagined script, sharing insights along the way, a woman stood up in the middle of the room, pointed her finger at me, and started to shout.

"*Why* should *we* listen to *you*?" she yelled.

I stopped midsentence, completely caught off guard. The room was silent for a few agonizing seconds. Then I tried to continue, explaining that one of our team members was a volunteer who had provided input every step of the way, when the woman interrupted again.

"What do *you* know about diving in that habitat? Have you ever done it?" she yelled.

A senior leader stood up and tried to come to my rescue, pleading with her to give me a chance. The tense dialogue between them got louder as I stood motionless at the microphone.

I had to get off that stage.

Moments later, I was hiding in the bathroom, hysterically crying. My pride and my passion were deeply hurt, and I was baffled. Why had she yelled at me? My intentions had been good. My only goal had been to lead this change effectively, creating and implementing the best possible show to bring value to both the organization and the visitors. What had I missed? Where had I gone wrong?

I had *assessed* and benchmarked where the project had started and where it could end up. I had *built* a plan and gathered a variety of stakeholder inputs along the way. I had *committed* to implementation. And I had *communicated* effectively—

No, I had not. I had missed the last part, and it was Communications 101: It's not enough to consider *what the message will be*. I had failed to consider *who was the best messenger* to deliver that message.

Yes, I was the team leader. Yes, I was passionate about the change I was leading. I was the driver behind every step—the person most committed to and, ultimately, responsible for seeing things through. But the target audience demanded credibility from their speaker, and while others had logged hundreds of hours diving in that tank, feeding fish, talking to visitors, juggling a thousand little details learned through time and experience, I had never even scuba dived in the tank. I could not directly relate to the divers' experience at all.

The best person to communicate the changes to that project's stakeholders would have been the volunteer on my team who could speak their language and answer their questions with authority. I had made a mistake in my first attempt at leading change—a mistake I have repeated a few times since then, although never to that level.

This leadership lesson has become deeply ingrained in my psyche and has a profound influence on my work today as I facilitate change with any business team or client. I am thankful for the insight it gave me early in my career and even for this audience member for so passionately pointing it out to me. I am a better leader of change because of this experience.

Cultivate Your Foundation to Withstand the Chaos

Just as the three species of mangroves work in conjunction to build and stabilize the shoreline, the three phases in this methodology will interweave to create a solid, cohesive yet flexible foundation for change. An unexpected diversion from a new colleague? A daunting new transition? An unforeseen burst of uncertainty? A shiny new opportunity to consider? Waves of newfangled ideas, uncertain paths, and murky, unforeseen challenges will always occur, but none of these can shake a solid foundation rooted in transparent planning and a systematic process.

I have used this simple methodology to broach new ideas with my team during a sixty-minute meeting and to guide clients through a visioning process over a three-year time frame. I've used it in my own work, to determine best next steps in building my business. I've used it to ensure thousands of stakeholders from around the world are equally and actively engaged in organizational outcomes.

If you are a left-brained, scientifically oriented person like me, this standardized methodology will help you craft direction and use rigor and structure to make decisions. If you are a right-brained, creative-minded person, this method will harness your energy into transparent, time-honored stages while also providing a concrete pathway forward so you can avoid the temptations of rabbit holes disguised as new ideas.

Regardless of how and when you choose to implement these three phases, rest assured the Mangrove Method will provide you with a solid foundation that can flex with you as a leader over time, as you evolve in your personal approach to lead and implement change. The three phases will work together like the three mangrove species to help anchor your shoreline as you face the waves of change.

The ocean can be as unpredictable as the next change coming around the corner—and both deserve (and demand) our respect and attention. Personally, I would much prefer to face a rogue wave among the comfort and stability of the mangroves.

Wouldn't you?

Unbreakable Law #1 Pro Tips

➤ When you feel overwhelmed by change, jump in the water and observe things from underneath for a new perspective on ideas and opportunities you may not have noticed before.

➤ To effectively lead change, use the Mangrove Method for a time-tested, systematic three-stage process—assess, build, and commit/communicate—that works together like the three species of mangroves to create a cohesive foundation and anchor your efforts.

➤ External forces may threaten your shoreline, but nothing can shake a solid foundation rooted in a transparent, systematic process for planning and implementing change.

Behind the Scenes with KiwiE

Standing in a desolate area in Hawaii, I listened to the scientist direct our group forward. Our destination: a rocky outcropping from which we could see the active volcano's lava flowing over the cliff and into the ocean. As the sun began to set, the glow of the lava became more prominent. Our excited chatter faded into the background as we walked along in silence.

But not quite silence . . . I started to notice the distinct yet unexpected sound of glass breaking. I looked around and saw nothing shaped like glass that I recognized, before glancing down at the black hardened lava under my feet. Through tiny cracks here and there, I could just glimpse the faint glow of orange hot lava flowing a few inches below. As I continued walking, I began to realize that the sound of glass breaking occurred each time my foot hit the ground.

To this day, I haven't forgotten that lava is predominantly made of silica,[3] because walking across the hardened lava field that day decades ago sounded like walking on top of a field of broken glass.

IF YOU'RE DISTRACTED BY FEAR, YOU'LL MISS THE OPPORTUNITY

(Spiders, Anyone?)

Working at a zoo, an aquarium, or a wildlife rehabilitation center—anywhere with animals, really—you quickly learn that animals, like people, do not always behave as we expect them to. Working with both animals and people together? Well, now you have two groups of living things interacting that bring daily challenges, changes, and opportunities to roll with the punches—because you never know what you are going to encounter.

One of the things I have enjoyed most throughout my career is helping people discover more about an animal or something in nature that scares them. I've seen people take chances and grow right before my eyes as they venture into the unknown. Often, they come back standing a little taller and with a spark in their eye. Fear is replaced by curiosity and wonder. Sometimes, though, this transformation takes more than one encounter.

Case in point: One morning my fellow aquarium educators and I were prepping to teach a class to local educators. We arrived early and were setting up when we noticed that the animal habitats in the back of the classroom were missing one resident—a snake. Because this snake was nonvenomous and accustomed to being handled by people (which is why we had it in the classroom in the first place), we weren't worried it would hurt someone. We just wanted to find the snake and put it back in its habitat.

As the teachers filed into the classroom, half-awake and yawning into their coffee cups, we realized that there were more participants than anticipated. I headed over to the corner of the room and began unstacking additional chairs, talking to my colleagues and the teachers around me. When I removed the second-to-last chair from the stack, there it was: a small, reddish brown creature coiled in a circle on the final chair.

"I found the snake!" I exclaimed, turning with a big grin to face the room.

Terror erupted as the teachers, now fully awake, screamed and scurried to the far side of the room. Afraid their cries would frighten the snake into slithering out of reach, I quickly reached for it and returned it to its habitat.

"It's just a harmless snake that lives in the classroom," I said, attempting to calm the teachers down. "Somehow it managed to escape overnight. All is well. We can now get started."

The teachers looked back at me with wary eyes. It took them awhile to focus on the subject at hand—and nobody sat near the back of the classroom that day. All I could think of was how much worse it could have been. If those teachers were afraid of snakes, imagine what would've happened if the missing creature was . . .

A spider.

Fear Is a Great Reminder

Change is scary.

Well, it can be.

Sometimes it's OK.

Actually, change can be pretty motivating—especially when it's your idea.

When you're the one with the vision, you can see the light at the end of the tunnel. When you're on the phone with the stakeholders and you control the budget, your hands are on the steering wheel and you feel confident about change. When you're excited to launch a new idea or try an innovative approach, your enthusiasm overrules your anxiety.

But when you receive an email out of the blue instructing you to change how you operate—or you go to a meeting and find out the way you've been doing your job is not how you're supposed to do it anymore—suddenly change doesn't seem so great. You may even be the type of person who digs in their heels and chooses to totally resist change. And I don't blame you.

But your instinctual fear of change doesn't have to be a cause for concern, and it doesn't necessarily hold up progress. Fear is simply nature's built-in "yellow light," reminding you to slow down—or even pause—so you don't skip over opportunities that might help you lead or participate in real change in your organization.

I'm not sure when it happens—probably at a different time for each of us—but somewhere along our journey to adulthood, we lose our innate ability to wonder. We no longer question without worrying what others will think of us; we wonder only whether we should have known the answer. The fear of not being seen as smart enough or "in the know" becomes a paralyzing force, and we retreat to the familiar. In the end, we lose our ability not just to marvel at the possibilities before us, but to even see those possibilities or our own potential—all for the sake of fear.

Honestly, I hope my mom never reads this chapter, because the mere mention of the word "spiders" sends her into a tailspin of terror. She loses the ability even to speak clearly. But now that I'm a mother myself, I sympathize with my parents on a whole new level. I think about the time when, as a child on a long-anticipated vacation to Hawaii, I found myself in a gorgeous condo on the beach, arguing with my younger sister about who got to sleep on the floor and who had to share a bed with our little brother. My poor jet-lagged mother and father attempted to quell our escalating conflict without waking our brother, already asleep in the bed. But my sister and I continued squabbling, oblivious to anything other than getting what we wanted—until we noticed my mother wasn't trying to stop us anymore. Instead, she was silent, frozen in place, staring up at the ceiling.

This unusual behavior jolted us back to reality. We followed her gaze up to the corner of the room, to the largest spider I had ever seen in my life. It was easily as large as my father's hand. After a brief moment of collective silence, my sister and I started screaming hysterically and jumped up on the kitchen counter while my dad, in a blur of parental bravery, chased the spider from the room.

Miraculously, the fifth member of our family, my brother, slept soundly on the bed through the whole ordeal.

My sister and I remained on the countertop, clinging to each other for dear life, long after the spider had left the room. You can guess what happened next: The argument magically transformed from who "got" to sleep on the floor by herself to who "had" to sleep on the floor by herself.

Sigh . . . my poor parents . . . and the poor spider that had innocently wandered into the wrong condominium. I'm sure somewhere inside my jumble of thoughts I was intensely curious about this spider: what it ate, how long it had lived to grow so big, what ate *it* in the end. To this day, I am always drawn to the spider exhibits. That was always one of my favorite parts of sleeping overnight with a group of kids at the zoo. Seeing the spiders after dark never seemed scary to me—maybe because I was so dialed in to teaching the kids more about these unusual creatures. But when I was a child on vacation in Hawaii, any natural sense of wonder percolating under the surface was immediately redirected as I watched my mother's instinctive reaction to the spider: sheer horror.

My career started with sharing with people the wonders of the natural world. It was impossible for me not to go to work excited each day about the joy I would share with others about wildlife and wild places. I've seen kids' faces when they jump into the ocean, open their mouths, and realize for the first time what saltwater *really* tastes like. I've seen teenagers' eyes pop when they wake up on a research vessel amid the broad expanse of the

Atlantic Ocean with no land in sight. I've watched the expression of horror in children turn into amazement when they take a chance and touch a snake. I've had teachers approach me with tears in their eyes after a student who hasn't spoken in their class the entire year speaks up to ask me a question about dolphins or sloths. Maybe a brief hesitation, maybe fear just below the surface, but all easily overcome by the magnetic pull of curiosity and wonder of animals and the outdoors.

Then we grow up and somehow, as adults, we're supposed to know it all and fear nothing. We're not supposed to question or give in to that natural wonder anymore. We're not supposed to admit our fear or rise up to the challenge of something new. We become far more accustomed to hiding our fears and continually *doing* than to seeing them as a yellow light to help us stop, wonder, and recognize an opportunity or the possibility of change.

I've had children run up to me at the aquarium, curious about the strange-looking animal swimming in the habitat right in front of us, only to have the parent interrupt and explain that it's a manatee and share all the associated facts they think they know. I don't have the heart to embarrass the parent and share that the animal is in fact *not* a manatee—instead it's a beluga whale. I can tell by the adult's face that they need to be right about this in the moment, so I let it go. But I've often thought: Why do they feel like they have to be right? Why are they afraid to ask me a question they don't have the answer to? Why can't they admit to their children that they just don't know and they wonder about some questions too?

Until I take my children on an outing to the local art museum. Or pick up my son, Kepler, from summer camp, where he was learning how to code. Or accompany my daughter, Tasman, on a program to learn about historical figures who influenced our town. Then, suddenly, I find myself in unfamiliar territory. Not only do I not know the answers, but I don't know how to even phrase the questions. My sense of wonder seems light-years away. And not knowing makes me fearful.

I realize at the same time that I should be modeling for my children how to feel fear, how to accept that you don't know the answers yet, how to ask the questions anyway, how to take a chance, to change, to grow, in spite of the fear. But ironically, this additional layer of fear—the fear of letting my children down—stops me in my tracks. I feel it even more deeply than the fear of not knowing.

I do empathize with people who don't know something and don't want to look like they don't know—because I've been there. When I'm leading an educational program or a leadership workshop, there are many such people in the audience—but, of course, that is the point of being there in the first place: to make a change, learn something new, take a chance, confront our fear of not knowing. Not trying out a new path or pausing to consider a new opportunity becomes a circular argument—we fear, so we don't question, so we fear—and impedes the natural *aha!* moments that could be magically transforming our fear into opportunity.

Ugh.

Fear has replaced opportunity.

We have lost our chance for meaningful change.

The spiders have taken over.

Perhaps you, like me, are highly motivated by change—the person who is always brainstorming new ways of doing things and new ideas to try. I'm a sponge for learning things and a machine for generating potentialities. I'm totally comfortable with change—when *I'm* driving it, or when it's happening in a familiar setting. I think I'm more afraid of *not* changing than I am of changing, so I strive full steam ahead.

Until I reach the point of something unknown.

When something is suddenly new, and I have to adjust—or admit that I don't know—then even I have to admit: The spiders of fear have taken over. I can't see through to the opportunities anymore.

Striving to always be up for a challenge and unafraid to rock the status quo might make you a challenging boss, but it can also make you a great leader. Depending on your team's comfort level with change and their willingness to step away from the "way things have always been," you'll either be a fantastic partner or their worst nightmare. I've been both.

But once you have rooted your plan for change in a solid foundation among the mangroves, you and your team can strive for the gray areas: the opportunities, the what-if's, the wonder. That's where the possibilities are, where the magic happens. Seeing fear as an opportunity waiting to happen can create space for new and wonderful things, in both your personal and your professional life.

During my favorite graduate course, called "Communicating Leadership," my professor said something that would become both the mantra for my professional work and my marching order in life: *Strive to make both yourself and your team feel safe, but not comfortable.* As leaders, we should work to create an environment where people feel safe enough to voice their opinions without fear and secure that they will be heard equally. But your team should not get too comfortable with the way things are, or they risk becoming complacent and just "phoning it in."

Innovation, creative sparks, and facing change head-on—all this can happen when people feel safe enough to get uncomfortable and imagine what could be.

Consider Change from All the Angles

Fast-forward a few decades from that family vacation in Hawaii. I was now living in Florida, on an acre of land nearly surrounded by woods. For Tasman, a toddler at the time, venturing outdoors and discovering all sorts of creatures living in our yard was a daily ritual. When Kepler was born, we introduced him to life outside too. Even before he could walk, I would tote him around in a baby carrier so he could feel the wind on his face and marvel at the commanding oak trees.

One afternoon, however, I experienced déjà vu when an animal from the outside world ventured into our home. Kepler and I had been out exploring and were returning to the house. I opened the front door and reached out my hand to turn on the lights—and found myself at eye level with a spider clinging to the wall, the size of which I had not seen since my childhood trip to Hawaii.

The spider and I stared at each other for a moment, neither of us moving. Then I casually glanced at my son and observed that the spider seemed to be about as big as his face. Suddenly, I saw things from my mother's perspective. I deeply understood in that moment how any sense of wonder and curiosity I might have about this spider could be instantly replaced by the sense of fear. Now that my role had shifted, my knee-jerk reaction of wanting to protect my son came spilling out so fast that I didn't even have a chance to intentionally stop and check in to see if my fear was warranted.

I took a breath and calmly shut the door to head back outside.

I would soon learn how misguided my fear had been.

Later, as I researched this enormous spider, I discovered what an amazing creature it is. Commonly called huntsman spiders, the species we have in Florida can grow up to five inches long and are extremely fast hunters. They do not build webs, so instead they rely on their speed and jumping ability to catch prey, often cockroaches. The female lays up to 200 eggs and protects her egg sac until all the spiderlings hatch. Huntsman spiders do not do well in cold weather, which is why this species may end up indoors during certain months.[1] All this was fascinating. And I am definitely on board with anything that eats a few more of our cockroaches here in Florida. I fully respected the spider as co-inhabitant of our land, and I accepted that it came with the territory when you live in a house in the forest. But the idea of a giant, super-fast, jumping spider that chases down its prey in my home? That brought even me to the edge. No disrespect to the spider, and I'm certainly not advocating for eliminating this creature. But as my mother immediately recognized decades earlier in Hawaii, I would be much happier if these awesome spiders would stay outdoors.

Leaders see a pathway forward that is full of possibility and ingenuity, or at least a necessary path of innovation that allows them to stay ahead of the fear (or the jumping spider). But all too often, their team can't see the same path. They don't trust the process of change. They fear the amount of work it will add to their "regular job." So, they dig in, give in to the fear of the unknown, and skip over the opportunity. Leaders must recognize both the fear and the opportunity. When not everyone on the team reacts positively to change, the leader must seize this chance to communicate productively so the team can move forward on the same page.

As an "ideas person" (which can be either a liberating fact or a terrifying reality for those around me), I naturally gravitate toward other ideas people, like my father, who can whip out a fabulous, interactive training program in no time. For people like us, the details and minutiae can become cumbersome and distracting. For example, I have been known to completely misplace my rental car—and even go to the wrong airport to catch a flight.

Whoops!

Thankfully, one person's spider is another's sea otter—in other words, what can be terrifying to some is fascinating to others. My fear of the details means I have learned to surround myself with detail-oriented people (see Unbreakable Law #5). Together, by combining our strengths and perspectives, we can approach opportunities for change without becoming sidelined by fear (eventually). I know this about myself now, but it's something I had to learn the hard way.

However, I've also come to learn how to become a detail-oriented person when I've needed to be. When no one has pulled up the map to figure out where we need to go and no one wants to call the customer to deliver difficult news and receive detailed feedback, someone has to. Without these details, or pathway forward, our work would start spiraling out of control. As much as I still abhor details, I've discovered a greater fear of wasting time and resources and dissolving into frustration. I've chosen to accept that overcoming my fear of inefficiency related to change is far more productive than digging my heels into who I've been until now.

As a result of choosing the opportunity to become more detail oriented when I need to be, I have found great success in my career as the middle person between people who are crazy-big-picture thinkers and those who are excellent at detail, execution, and getting it done. And the hierarchy we represent isn't always how you might imagine it. Sometimes the leaders above me have been the big-picture thinkers, but sometimes the bird's-eye-view possibilities come from my frontline team members. Either way, often it has been my job to communicate this 30,000-foot view up or down the food chain, while also removing any obstacles to change.

The most common obstacle, of course, is fear.

On one occasion, I was leading a large project that had been many months in the making. The timeline was set, communication of the plan was ongoing to multiple departments, and the people responsible had been activated to officially start the project the following Monday. Then, on late Friday afternoon before "go day," I received a phone call from another department head.

Up until this point I had heard nothing from this leader. Now, however, he explained his hesitations on the project and how it would affect his team. He detailed the negative impacts not only on his team's performance, but also on the deliverables by which their progress would be measured. He was distressed and adamant.

From my perspective, I had two options.

Option #1: react. My initial natural reaction was to unleash my own frustration on the department leader. Why was he bringing up these ideas at the last minute when he'd had months to do so? By choosing to voice his worry about his own team at this late hour, he was negatively impacting my team, in terms of both their deliverables and their positive mindset going into this big change on Monday. That was what I *wanted* to say.

Option #2: listen. After a moment of internal fuming, I wondered how I could think creatively about this sudden development. How could I assuage his concerns without acquiescing completely? How could I acknowledge the fears he brought to the table about the impending change? Even at this late hour, his concerns were valid, and I could at least lend them an ear. I still had my team's back, and I wasn't about to undermine the work they had put in. But perhaps I could build a critical bridge with this other department—a bridge that could benefit all of us, now and in the future.

I took a deep breath and chose Option #2.

I listened. Then he listened. We brainstormed ideas. Neither of us were completely satisfied, and both were a little frustrated. The change would still go forward as planned on Monday, but with minor accommodations and observations on how our teams might collaborate better in the future. Keeping in mind that the implementation would affect the entire company, we agreed to keep the dialogue going.

Then I had to tell my team.

It did not go over well.

To be fair, it was late on Friday afternoon and my whole team was exhausted. They had put their heart and soul into the details of this transformation. They were done talking, done listening, and ready to execute. Receiving input at the last minute, when they had been requesting it for months, was not part of their plan.

I listened to my team. I heard the fear and frustration in their escalating voices. I watched them shake with anger, and I knew they were bone-tired. And I agreed with everything they said. It was everything I would have said—and probably did say—when I was in their role. But now, as the leader, it was my job to help them look at the big picture.

Yes, we were the champions of this change. We were the people responsible for making it happen and for the associated outcomes—yet it would not affect us alone. We had to keep the broader team, the company as a whole, in mind. I had to help my team see things from another angle rather than become distracted by the fear of little changes. We could adjust our overall plan to build a collaborative relationship with an integral department within the company.

I asked my team to envision this as a sacrifice for the greater good, for the viability of future changes our team would want to make. It was not easy, nor were they happy about it. In fact, they were afraid. Would all their hard work be undone? Would these minor adjustments derail the overall project?

I stood by them and applauded their herculean efforts in the final hour, appreciating that they employed these efforts through sheer trust in me. They still could not clearly see the big picture—but it wasn't their job to see it. It was my job as their leader, and they trusted me enough to implement change even while fearing the fear. This is what my graduate professor had meant about the possibilities when we feel safe, but not comfortable.

The other department head noticed the efforts our team made. He communicated his impressions to his team so they would know that their colleagues heard their concerns. New professional respect was formed between our teams, which served us well when even more challenging ideas and changes came our way soon afterward. This willingness to see things from all angles created a lasting legacy that I am extremely proud of to this day.

What opportunities had fear almost distracted us from recognizing? There were many: the opportunity to understand colleagues who previously had not been very forthcoming. The opportunity to notice original ideas and try things in a new way with input from a different team. The opportunity to discover our ability to flex and adjust, even at the last minute. The opportunity to make the project even stronger. And, for me personally, the opportunity to grow as a leader whose team trusted me with the unknown.

Respect Your Fear and Uncover Its Purpose

Do all opportunities pay off? Of course not.

In the work scenario described previously, many things could have gone wrong. The last-minute changes could have totally derailed our project. Making the minor accommodations could have driven a wedge between our two teams. My team could have revolted and chosen not to implement the ideas I suggested. The new measures could have utterly failed, and everyone could have lost trust in me as a leader. All these are catastrophic possibilities—but the possibility for catastrophe always exists, whether you plan for a change far in advance or you are pivoting at the last minute.

Fear exists for a reason. The autonomous "fight or flight" reaction—in people and animals alike—is real, and it has served us well over time. It is wise to pay attention to fear and to listen to your sixth sense. The hairs that stand up on the back of your head do so in order to tell you something important. Fear is healthy. It helps us survive.

I love spiders. I find them fascinating. I am grateful for their contributions as a part of the ecosystem and in awe at the wonderous ways they have adapted to their various surroundings. But if I come around a corner and unexpectedly encounter a spider, I still jump.

Can fear distract you from opportunity? Yes.

Should it? Sometimes.

Can you feel the fear and move forward anyway? Yep.

The first time a Chilean rose hair tarantula was placed on my open palm, I tensed. I might even have shivered ever so subtly as I watched her extend her chelicerae and clean her fangs. I will admit to feeling slightly vulnerable as I held this incredible creature and watched her move. I was learning how to handle the zoo's ambassador animals, and having a healthy awareness of any fear I—or my audiences—may have was not only advisable, but also an essential part of the job.

Ever felt slightly vulnerable when you're handed a new opportunity to effect change? Tensed up before you led your first meeting on this change? Felt a little exposed when you laid all of your great ideas out for feedback?

Did all this fear serve a purpose for you? Could it have made you stop and intentionally include a wide range of opinions rather than relying on your own or a few close contacts? Could fear have caused you to take extra time preparing for the meeting rather than doing it the way you always have? Could fear have made sure you deliberately asked questions to receive feedback—questions you really wanted to know the answers to—rather than asking questions out of routine?

You are learning how to lead change at every point in your career because the next change you lead will always be (at least) slightly different than the last. Having a healthy respect for your—and your team's—fear will ensure you pay attention to that yellow light and let it purposefully direct your attention to things you might have missed.

More recently, I visited a zoo during Halloween season with my children. The creative seasonal decorations added a fun aspect to our experience. My son ran ahead, climbing on every possible structure, while my daughter lagged behind, reading all the signs, particularly amazed at the graphic explaining how they took care of the new baby elephant born at the zoo.

We rounded a corner and ahead ran my son, per usual. I kept one eye on him as he zigzagged through the crowds before coming to an abrupt stop in front of a giant inflatable spider. The silly, enlarged eyes made him laugh. Then he turned left and headed into a tunnel-like structure, its metal arches enclosed with see-through mesh. It looked like a typical butterfly habitat.

I called ahead to Kepler, urging him to rejoin us on the trail through the zoo. I wasn't particularly interested in seeing butterflies that day. But he called back, insisting that I join him instead. "Mom! Come on! This is really cool!" he exclaimed.

Annoyed at this detour, I tried again. "Honey, we need to keep moving toward the giraffes. We have to meet our friends, and the rain is coming soon—"

"Mom, you *need* to come here." He was having none of it. "I promise."

The tone of his voice stopped my automatic reply. He was not one to argue. And sometimes my kids seem to know me better than I know myself.

I was sure this was a regular old butterfly exhibit rather than anything particularly interesting. I do love butterflies, but I was more interested in seeing the giraffes that day. Surely the inflatable spider was just another decoration for Halloween.

I could not have been more wrong.

What I walked into was indeed structured like a typical free-flying butterfly habitat. The creatures were all around me—at my head, my feet, and everywhere in between. I had to be careful that the chain-link curtain closed behind me so no animals would escape. I had to watch where I stepped so as not to crush them and watch where I put my hands lest I injure one unintentionally.

But there were no butterflies in this exhibit.

These were spiders.

Hundreds of them, all around us.

Over our heads and next to our faces.

Straight in front of us as we inched ahead, and right behind us when we turned around.

Next to us on all sides—and yes, potentially underneath our feet if we didn't stop to look before we stepped.

I have visited more than sixty zoos and aquariums, in the U.S., Canada, Australia, New Zealand, and throughout Europe, and I have never seen an exhibit like this. My first instinct was to be terrified. But then my inner child's wonder took over.

I was mesmerized.

I adored being right next to these spiders, up close and personal, watching as they went about their spider activities. As I turned my head from side to side, I could see spider legs mere inches from my eye. I watched as some ate their webs.[2] Others devoured prey. Still others sat motionless in their webs. One to my right was meticulously weaving its web, and just overhead one dangled from a thread almost unseen to the naked eye.

I had never been surrounded by spiders like this before. I didn't want to leave. I could have stayed in there all day and then slept overnight just to watch how their behavior changed with each passing hour. My children had to call me to leave—a couple of times. Eventually, I tore myself away. But the giraffes now were no competition for the spiders.

Change is scary—potentially as scary as spiders. It can be overwhelming, intimidating, exhausting. It can seem to defeat you before you've even had a chance to say "go."

But change also can be fascinating. It can be totally different from what you expected. Perhaps it will even be the most liberating, energizing, magnetic experience you've had in a long time.

What's the point?

Every change brings both opportunity and fear, on a sliding scale with many variables. How big the opportunity is, and how overwhelming the fear can be, is determined by the particular change—and by how you and your team react.

Fear may be real and warranted, or it may be a distraction. Look it square in the eye (or multiple eyes, in the case of eight-legged fears), and figure out which is which.

Then leverage the fear.

Embrace the opportunity.

Remember the spiders.

Unbreakable Law #2 Pro Tips

➤ Fear is nature's built-in yellow light, reminding you to slow down and pay attention to the opportunities for leading change that sticks.

➤ One person's spider is another person's sea otter, so use your team's complex reactions to change as a chance to level the playing field for productive communication and progress.

➤ Although fear (like spiders) may seem strange and harmful, it always serves a purpose—and it deserves our attention and respect.

Behind the Scenes with KiwiE

Arriving one morning for work at the zoo, I heard a buzz of commotion: Our young tigers were going to be walked throughout behind-the-scenes areas that day, so they could get used to the smells, sights, and sounds of their new surroundings. Sure enough, two trainers were soon walking down the hallway with two beautiful animals alongside them.

I smiled along with the rest of the team, thrilled for the chance to watch the tigers up close. Then I turned back into my office to start work for the day . . . and stopped in my tracks. There in my office were a turtle and a sloth, accompanied by their own trainers.

I shot a quizzical look at my teammates, who explained, "We didn't think it was smart to put too many distractions around the tigers just yet." Nodding my head in agreement, I turned to pick up the ringing phone on my desk. On the other end, a friend was calling to ask how my day was going.

"Well," I replied, "I'm getting ready for a few meetings, there is a sloth and a turtle in my office, and two tigers are walking down the hall outside my door. Typical Monday morning at the zoo. How's *your* day?"

An extended silence was followed by, "Ummm . . . I really don't know what to say to that."

WHEN YOU CAN'T SEE THE FINISH LINE, LET PURPOSE BE YOUR GUIDE

(Sea Turtles for the Win!)

Ask me which animal is my favorite, and the answer will be instantaneous: a wolf.

I've gone out of my way to visit zoos and rescue centers that are working diligently to protect and learn more about wolves. I've cheered from afar as they were reintroduced back into Yellowstone National Park. I've driven almost as far north as you can get in Minnesota, to a little town called Ely, to visit the International Wolf Center. I took a winter ecology graduate course at the Teton Science School in Jackson, Wyoming, specifically so I could spend time outdoors in below-freezing weather, waiting for hours on end in one spot while observing elk behavior, on the off chance I might catch a glimpse of a wolf. (I didn't.)

One day I will see a wolf in the wild. It's the top wildlife goal on my bucket list. But when I stop and think about it, I realize that sea turtles have played an even more prominent role in my life than wolves. Back before I even fully realized how amazing they are, I unintentionally relied on sea turtles to help me figure out what my next step in my career would be.

At the time, I couldn't see the opportunities ahead. I was blinded by my love of my current professional role, my friends, and my city. I was terrified of leaving behind everything I had worked so hard for up until that point. I wasn't sure I wanted to take a step out into the unknown.

Leaving my dream job at the aquarium was the hardest decision I've ever made in my career, but something was calling me to make my next move. Perhaps it was the experience I'd gain by moving from a department of thirty people to a department of three. Or maybe the chance to be in the midst of groundbreaking scientific research in real time. But the answer I gave the most often when people asked why I was leaving?

I wanted to watch as a sea turtle laid her nest.

I wanted to observe the female coming out of the water at night, crawling up the beach, choosing a suitable spot to lay her eggs. I wanted to know if sea turtles really do have bioluminescent organisms living on their shells that I could see glowing in the dark at night. I wondered what scientists must experience when they approach a female and place a tag on her flipper so they can track her movements and learn more about her life in the sea. I wanted to be *right there* as scientific discoveries were unfolding. I wanted to share in the findings and the excitement.

I didn't know what my finish line would be or how I would deal with this life upheaval, but I could sense this was the right step. So, I took a big breath and—even though I could not see the end in sight or really understand what my final career or life goal was—walked forward into the unknown.

Focus on Finishing No Matter What

One of the hardest parts about change is never feeling like it's over. Do you remember ever leading your team through a change that seemed like it went on forever? Or maybe you feel this way right now—as though every time you get to a milestone, rather than being free to step back and celebrate how far you and the team have come, you face another change handed down from above.

The prospect of a "finish line"—the idea that there is an *end point* to whatever you may be working on, and a chance to rest and reflect before beginning again—is familiar and motivating, whether in your personal or professional life. Maybe you are a runner and have actually crossed countless finish lines in your life. Maybe you are a musician, and your version of a finish line is the last note of a song you have been working on, or the thunderous applause of the audience appreciating your performance. Maybe you are an engineer, and your finish line is the last bolt tightened into place before you stand back and admire your creation brought to life. Maybe you are an author, and your finish line is when you send your manuscript off for publication after months, or even years, of writing.

In the business world, a finish line can take the form of anything from the completion of a capital campaign that successfully raises millions of dollars for a building's new wing to transitioning your team away from in-person interactions as they adopt virtual methods of working. The finish line can seem—and may actually be—miles down the road.

When change is happening—whether you're leading it, you're participating in it, or it's happening "to" you—it can feel both exciting and overwhelming. No matter which side of the coin you are on, the universal feedback I receive about change is that people want to know when it will end. Whether they're hoping to celebrate a big success or simply that the changes have ceased, they want to see the finish line. At the very least, they want to *know* that they are close to the end.

Change may sometimes feel difficult to manage or lead, as if you are moving as fast as you can through heavy sand and not making much progress. But there is always an end point, no matter how far away or unattainable it may seem in the moment. Until you reach it, you'll better serve yourself and your team by both remaining focused on the finish line up ahead, and by achieving the mini-milestones of continually placing one foot in front of the other.

Remember the story about the tortoise and the hare? The hare ran as fast as he could, straight out of the gate, but then slowly his energy waned and he decided to take a nap. Meanwhile, the tortoise, slow but steady, ended up crossing the finish line before the hare. By keeping an eye on the prize and never giving up, the tortoise came up with the win.

Consider the oceanic relative of the tortoise, the sea turtle. These creatures have existed for more than 100 million years—since the time of dinosaurs.[1] And from the outside looking in, a sea turtle's way of life seems not just mind-boggling, but virtually impossible.

A female sea turtle swims on a journey of many thousands of miles on her way to a specific nesting location. Once there, under cover of night, she will leave the ocean and begin to crawl slowly but steadily across the sand. Every few minutes she will pause, take a deep breath, then continue searching for an appropriate spot to lay her clutch of eggs. Then she uses her back flippers to dig a hole a few feet deep and deposits up to 100 eggs. It may take her two hours or more to finish this nesting process— after swimming thousands of miles through the sea. And she does all this without GPS, without doulas, without midwives, without any external guidance. She relies solely on her instinctual compass within, guided by a singular purpose.[2]

I once worked with a manufacturing client that expressed frustration about a training program used to bring new staff up to speed. The client explained that new hires are placed on rotation with experienced team members who can show them the ropes, how to use the equipment and the overall job responsibilities. After shadowing the veteran team members for up to two weeks, they are expected to pass a test that will qualify them to operate the equipment on their own so they can be placed into equal rotation with the other employees.

This was how it had always worked, but now they were hearing push-back and frustration from the new hires. The program relied on the staff person's ability to self-direct and take initiative in addition to being directly taught and guided. Now, this was not proving successful. The leadership didn't understand what had changed.

Listening to my clients explain their quandary, I casually observed our surroundings. We were sitting in a training room in the middle of a large manufacturing plant that was mostly outdoors. We were all wearing company-appointed personal protection equipment (PPE) and following appropriate safety protocols. Just that morning I had been briefed on the latest safety measures, had signed in on two different visitor logs, and had been made aware of recent visits by federal safety and health inspectors. I was escorted everywhere I went and carried the certified safety card I'd earned while attending the updated safety briefings.

The equipment surrounding us was, in some cases, hundreds of feet high and maintained by highly complicated electrical systems. An intricately coordinated process involving fire, water, materials, and more combined to make everything work. Well-timed interactivity between various departments ensured everything was functioning properly, producing product on schedule, ready for scheduled maintenance to minimize downtime, and adhering to strict environmental standards.

I had been working with this company for years and was dialed in to company standards for an external consultant, and I still learned something new every time I was on-site. And they were wondering why a newly hired team member wouldn't feel ready to operate intricate, expensive, *combustible* equipment within the first two weeks on the job.

The answer was not in any unrealistic expectations. The company had been operating this way for years. The answer was in the outdated training methods they were using—an unreasonable portrayal of the finish line that new hires were expected to cross.

Instead of throwing everything at the new employee and then expecting success, we broke it down into digestible components that could be taught, understood, implemented, and self-directed over a reasonable time frame. I worked with the lead trainers and mentors to help them better understand current methodology on how to teach and how people learn. Together, we created a system with smaller, more manageable milestones along the way, leading up to the big finish line. What before was a herculean task of learning how to successfully operate impressively powerful machinery without breaking it or blowing it up, now became an exercise in small steps toward overall progress that both the trainer and the trainee could implement on their own.

The finish line that had once seemed so far away—and almost impossible to see through miles of murky water and unclear direction—became an achievable goal with breaks and recognition for milestones achieved along the way. A detailed, digestible training program provided an instinctual compass for each new employee to navigate their way, just as the sea turtle does.

Trust Your Instinct Every Step of the Way

No matter what kind of finish line you are working toward, in any aspect of your life, one truth remains fundamental: Bumps in the road will appear, some planned for and some unexpected, and you must adjust and adapt to keep moving along your planned course. And as you do, there may come a point when you feel sure you should be reaching the end, when everything should be falling into place, yet you still can't see it in reality.

No one ever said change was easy.

I know that as a leader you are fully dedicated to reaching that finish line with your team. But when those bumps in the road arise, and when there's no end in sight, often you might very much like to lean on others around you—until you notice they are just as tired as you are and just as ready to reach the finish line. That's when you must dig deep, trust your preparation, and be guided by your own instinct so you can truly lead from the inside out.

Remember the sea turtle, who stuck to her goal and completed her mission? Well, here's the rest of the story: After laying her eggs, she covers the nest with sand, makes her way back to the ocean, and swims away into the dark of night, never to see her hatchlings again!

That's right—the tiny sea turtles are left to fend for themselves in the wild. A few months later, usually at night, the eggs will begin to hatch in the nest, and the sea turtles will crawl over one another, aiming toward the surface above. In an event aptly described as a "boil," up to 100 hatchlings seem to suddenly erupt from beneath the sand and head toward the water. They crawl as fast as their little flippers will take them—avoiding ghost crabs and other predators, as well as debris on the beach—until they reach the ocean. Each slogging on its own way through the sand and into the sea, where they will face even more challenges. The chances of a baby sea turtle surviving to adulthood range from 1 in 1,000 to as high as 10,000.[3]

In the seawater, the baby sea turtles are a bit less unwieldy. Again, they rely on their little flippers, to swim as fast as they can, avoiding all sorts of traps such as fish that might eat them and floating plastic that might entangle them, until they reach safety: a mass of seaweed floating in the sea. There they will grow into juveniles before starting their lives as adults, potentially thousands of miles from where they hatched. And if they survive, how do they manage all this? One word: instinct.

No one ever said being a baby sea turtle was easy.

When change feels the hardest, and when you may feel the most alone, is the exact moment when you need to trust your training and gut instinct as a leader. The road of change is not straight. It will feel like you are slogging your way through heavy sand. You cannot always see the end. There will be twists and turns. There will be potholes and ghost crabs ahead. You will even notice places where you are tempted to make a U-turn.

You may be surrounded by people along this road, or you may reach a point where you are all by yourself. You may have a mentor to reach out to, or you may be on your own just like a sea turtle hatchling. Whatever the circumstances, it will take courage, tenacity, and focus to keep going along this road of change. You will have to dig deep, trust, and at times cling to your leadership instincts.

And finally, you will have to hold tight to the purpose for which you started down the road in the first place.

Hold Tight to Your Purpose

Here is one of the things about sea turtles that amazes and impresses me most: When it is nearly time to lay her eggs, the female sea turtle finds her way back to the region from which she was hatched. She migrates using Earth's magnetic fields and other cues deeply ingrained in her instinct, until she arrives at the place (or very nearby) where she herself emerged from her egg in a nest deep under the sand. She repeats the process of her ancestors, crawling out of the ocean onto the beach and laying her clutch of eggs— the next generation of sea turtles. She does all this without "seeing" where she is going, without "knowing" where her finish line is. Instead, she relies on deeply instinctual and evolutionary migratory clues to guide her way.

She is guided by purpose.

What is your purpose?

Why did you start along this journey of change? Was it so you could increase operational efficiency, which will positively affect the organization's financial standing this quarter?

Why did you decide to start weekly team-building sessions? Was it so you and your team could work to better understand one another as individuals and professionals, because you have a tough business outlook in the months ahead?

Why did you choose to chart a new course for your professional life? Was it so you could build a new skill set or impact the world in a different way?

I've experienced various purposes like these in countless iterations, in both my personal and professional life. I've been a part of cheerful teams that were totally on the same page to face change head-on and implemented new methods of operation in record time to overwhelming accolades. But I've also been the lone wolf, questioning our approach to the change, bringing in unpopular opinions that slowed down the forward progress.

I've been the person excitedly leading the change, so sure the finish line is around the corner that I skip ahead to the end—which sometimes results in disastrous consequences. And I've been the grumpy leader in total disagreement with the course of action, anxious to get to the finish line to stop being distracted by time-wasting efforts.

Each time, I have felt the magnetic pull of purpose guiding my actions. I've had to listen to that inner voice telling me to speak up and question a decision I disagree with, if I thought it was vitally important to the change we were working to implement. I've had to show up and motivate people—including myself—even when I was a bit overwhelmed in the moment, if it meant we kept moving forward along our course. I've had to politely decline a project offered or a chance to weigh in, if I felt the addition would tear me away from the focus on the change I was currently leading or even a new change I wanted to tackle.

And all along, I've innately understood the concept of a "finish line"—even if it felt more like a nebulous concept in my head than an actual finish line in reality. I knew that the process of change must come to a conclusion at some point in order to measure progress and recognize our efforts. I've been guided by professional purpose, propelled toward an imaginary finish line, until, after sixteen years of career experience, I found myself in a totally unfamiliar and intimidating space with an *actual, real-life finish line.* At the time, it seemed to have nothing to do with my career goals or aspirations. But as I would soon learn, this new journey would teach me more about purpose and the process of change than I could ever anticipate.

I decided to complete a marathon.

If you're anything like me, you have gone most of your life with a distant appreciation for marathon runners—as in, *Wow, I can't believe all those people are really running . . . how far is a marathon again? Wait, 26.2 miles?!* Running was something I did to warm up before playing tennis or during boot camp fitness classes, but I was never one to just "go out for a run."

I really *wanted* to be. I envied people who could run for miles without stopping or seeming to break a sweat. Living in downtown Chicago, I was excited to cheer on the marathoners every October, and I challenged myself to a 5K race every now and then. But the concept of being "a runner" was foreign and out of reach.

Until I became a mother.

When my two children were born eighteen months apart, I suddenly found myself yearning for a return to my usual physical fitness, caring for two babies simultaneously while balancing every other aspect of my life, not to mention the aftermath of a complicated birth: massive scar tissue, a life-threatening infection, chronic muscular instability, searing pain when just walking across a room. One evening, as I sat at the computer drinking a glass of wine and eating a brownie (#truth), I came across a random email inviting me to participate in a training program. The goal: to walk a half-marathon.

I put down my glass of wine and decided that this challenge was my new plan to get healthy in mind, body, and spirit. The idea of having time to myself while walking for miles on end (which would take hours!) was totally appealing—a new mother's dream.

I took the leap and signed up.

Months later, tears welled up in my eyes as my training partner and I approached the end of our first half-marathon. We were slow—the pacing truck was not far behind us—but we never stopped moving forward—not until we proudly walked under the giant blue, inflatable archway labeled FINISH LINE.

I hugged my children, exhaled into the moment, and my first thought was *Hmm, I actually made it . . . I finished 13.1 miles!*

My next thought? *If I could do that . . . maybe I could complete a whole marathon.*

Fast-forward a few years, and I have completed twenty-one half-marathons and five full marathons. I have learned not only to become a distance runner, but to keep going along an uncertain road when you literally can't see the finish line up ahead.

For my inaugural marathon, I decided that if I was going to endure running 26.2 miles, I might as well do it at the happiest place on earth: Disney World. Surely if I got tired or nervous, the costumed characters would keep me going.

When you run the Disney Marathon, you wake up at 2:00 a.m., dress in your Minnie Mouse (or other character) finest, and take the bus to EPCOT, where you join 25,000-plus new running friends at the starting corrals. As the race begins at 5:30 a.m., each wave of runners is released with a fireworks display and the Fairy Godmother gleefully waving pixie dust over us.

The course travels through all four Disney Parks: Magic Kingdom, Animal Kingdom, Hollywood Studios, and finally back into EPCOT. You can take a break anywhere along the way to hop on a ride or get your picture taken with characters such as the Haunted Mansion ghosts. After the first hour of running in the dark, you suddenly find yourself running down Main Street in the Magic Kingdom, with thousands of people lining the street screaming for you, including Buzz Lightyear, Sleeping Beauty, Prince Charming, and Alice in Wonderland. Around the next corner, trumpet players on a balcony announce your arrival as you run through Cinderella's castle.

A few hours later when you reach EPCOT, the sun is shining high in the sky and random park guests are cheering you on, offering water bottles, ice cream, and even beer. During the final one-tenth of a mile, a gospel choir is singing for you as you break into an uncontrollable grin. A few moments later, you cross the finish line and give Mickey Mouse a high five!

My second marathon could not have been more different than my first: I chose to run the Marine Corps Marathon in Washington, D.C. I wanted to run on hallowed ground and honor our history as a nation, following in the footsteps of people who have sacrificed for us all. "The People's Marathon," as it is known, also offered me an opportunity to fundraise for a worthy cause. Instead of fireworks and the Fairy Godmother cheerfully sending us on our way, a Marine welcomed us by shouting into a microphone as Medal of Honor recipients parachuted into the starting area to run the race with us. A four-star general walked by directly in front of me and runners on all sides stood at attention.

It was a cool October day, which was perfect for running, and I started the race full of excitement and confidence. Instead of theme parks and animated movie characters, I was running past national monuments and alongside people carrying folded flags in honor of their loved ones. Mile 13, known as the "Mile to Remember," was a long, straight stretch with people on both sides reverently waving American flags in honor of the servicemen and servicewomen who paid the ultimate price in service to our country. And each water station along the course was staffed by volunteers and active-duty Marines yelling, "Come on, runners! Oorah!"

Again and again, I gratefully accepted their commanding encouragement while tearfully blubbering, "Thank you for your service!"

I was thoroughly enjoying this race . . . the beautiful day . . . the victory lap after all the training . . . when suddenly, at Mile 19, the wheels started to come off. My legs seized. My stomach churned. I needed to slow down. I grabbed an energy bar and drank some water. The world melted away around me, and my field of vision narrowed. I became entirely focused on continuing to move forward no matter what. I knew I had to get to Mile 20 if I had a chance of finishing this race.

Just keep moving forward.

Cross the last bridge, then you're on the home stretch.

Remember why you started.

It was one of those moments when you realize that the only thing you can do is trust your training. What I love most about running marathons is that what you put in to get to this moment is exactly what will take you to the finish line. You cannot cut corners. You cannot talk, schmooze, or cajole your way through. There is no "fake it till you make it" in a marathon. Running 26.2 miles demands that you show up ready for the whole journey if you want to reach the finish line.

And whether you are an elite runner or a weekend warrior like me, every single person is challenged during the final stretch. Your body endures physical changes, literally fueling your motion differently from the previous twenty miles. Your legs start to burn, and your mind questions everything. That's when you lean as hard as you can into your heart and soul, remembering *why* you started running this race in the first place. You still can't see the finish line yet, so holding tight to that purpose will be the only way you'll make it there.

I caught a glimpse of the Mile 20 marker up ahead, and I could feel myself start to relax. I snapped a picture with the Marine at the water station and picked up the pace, winding along the next few miles parallel to a highway. I began to envision Arlington National Cemetery, which would signal the end of the course. I found an inspirational song on my iPod and put it on repeat. And I dug down deep within.

I started running with the goal of getting healthy in mind, body, and spirit after the birth of my babies—and so I could model for them how to be a healthy individual.

I chose this race because it is two weeks before I turn forty, and I wanted to prove to myself I could do it.

I'm running in gratitude for all the members of my family, friends, and community who were and are part of our military.

And I'm running in one of the birthplaces of American history, to celebrate this beautiful country that I am so proud to be a part of.

When I could finally see the Mile 26 marker up ahead, I aimed straight for it, with a massive smile on my face. I reached the marker, went to take the next step, but in the final two-tenths of a mile of this massive race—I suddenly stopped dead in my tracks.

Throngs of screaming supporters surrounded me on all sides. The tall grandstands, full of spectators, marked the hairpin turn toward the finish. Music was blaring, and an announcer was calling out the names of the runners who were crossing the finish line. It was unbridled energy mixed with chaos in every direction. All the forward movement felt like it might just take me with it toward the end. I was so close, just seconds from finishing the race, after hours and hours of running. But I couldn't actually see the finish line yet.

There I was, standing still right in the middle of the course, totally spent in every sense of the word. I had absolutely nothing to give—no well of reserves to dig from anymore. I. Could. Not. Keep. Going.

A woman stepped from the screaming crowd and onto the edge of the racecourse. I had never seen her before, and never would again, but she looked straight into my eyes and calmly but firmly said, "You can do it."

"No, no, I can't," I sobbed. "I have nothing left. I can't keep going. I don't even know where the finish line is."

"Yes, you can. You can do it," she said again. "The finish line is right around the corner. *Run.*"

And she was gone.

I turned back into the course, on autopilot now, and leaned into the pack of runners. And with everything I had left, I ran.

I have no memory of actually crossing the finish line—no recollection of what it looked like, or a feeling of euphoria, or jumping in celebration. I just know that I tucked my head and focused solely on the next step, and the one after that, and the one after that. I remember getting to the top of that gradual hill, turning right, and running and running and running . . .

I had no idea I had even finished the race, until a Marine held out his hand to give me a high five, looked me in the eyes, and said, "Job well done." I glanced ahead and saw a line of active-duty Marines as far as the eye could see, all waiting to also give me and the other finishers a celebratory high five.

With each slap against my hand, I broke more and more out of my trance, and tears started rolling down my face. I followed this line until somehow I found myself standing in front of a Marine, directly in front of the United States Marine Corps War Memorial, as she put the medal around my neck.

They say you run a marathon in three segments: first with your legs, then with your head, and finally with your heart. That last segment is what typically separates the people who finish from the people who don't. I have found this to be true in my journeys as a marathoner, but also in all aspects of life that take strategic thinking, epic preparation, diligent follow-through, and continual recommitment to your core purpose. And I believe every journey though change embodies these components.

Just like the sea turtle's purpose leads her back to the same beach she came from, so your purpose as a leader can guide you through what might feel like thousands of miles of change.

Just like the baby sea turtles hatching on the beach, with no adults in sight to teach or lead them, you can operate on the most profound training a leader can access: instinct.

And when things inevitably get tough, keeping your focus on what lies ahead—no matter what, even if you can't quite see it yet—will get you through to the finish line.

Every. Single. Time.

Unbreakable Law #3 Pro Tips

➤ Change can feel like making slow progress through heavy sand, but keep your focus on the finish line, no matter how far away it seems.

➤ When change feels the hardest, trust your training and instinct as a leader, remembering the baby sea turtle that heads toward the ocean with confidence, trusting its instinct every (flipper) step of the way.

➤ Power through the uncomfortable parts of leading change—when you feel all alone in the middle of the ocean, swimming toward an end you can't quite see—by letting the magnetic field of your purpose guide you.

Change is constant, but it doesn't have to be chaos.
If you're distracted by fear, you'll miss the opportunity.
And when you can't see the finish line, let purpose be your guide.

You can do it.
You can lead change.
Keep running toward the finish line.

Behind the Scenes with KiwiE

Overnight experiences at the aquarium were always a big hit for visiting student groups. I always loved sleeping in the underwater viewing areas too, watching the behavior of the animals change as day turned to night and back again, through the eyes of children.

One morning, we awoke to see sunlight streaming into the habitat, a large tank filled with manatees and all kinds of freshwater fish that was open at the top, far above our heads, to the outside world. Around the room, I could hear yawns turn into *oohs* and *ahhs* as the kids remembered where they were and felt the excitement of being where not many people get the chance to be. A group of kids were especially excited about seeing animals swimming right in front of where they had been sleeping.

I was sleepily executing my morning chores of putting sleeping mats away, getting the educational activity ready, and setting out breakfast, when I heard:

"Ms. Julie! Ms. Julie! It's an *alligator!!*"

I yawned and continued my work with my back to the habitat, not looking up. "I'm so proud that you are practicing the observational skills I taught you last night. You're right—the alligator gar is a fish that looks very much like an alligator."

"No," they exclaimed. "It's a *real alligator!*"

"Yes," I calmly responded, "this fish *does* look like an alligator. But it is a fish. Can you observe what they have in common?"

The kids, not to be deterred, began recruiting other classmates. Their chorus of voices grew louder and I kept responding with my same, monotone, half-asleep answer.

"YES," I answered again, now somewhat exasperated, as the fifth group of students came rushing up to me, chattering on about the alligator. "I know this fish does *look like* an alligator, but it *is not* an alligator. It is a FISH."

Then, as I had finished my morning responsibilities, I turned around.

There, swimming happily among the other animals, alongside the alligator gar, was—an alligator. It must have crawled into the tank overnight and was now apparently content to enjoy our native animal habitat.

I turned back to the kids, now swarmed around me, waiting with eager faces. "Yes," I admitted, to their delight. "You are right, and I was wrong. That is indeed an actual alligator."

INSTINCTUAL LEADERSHIP FIELD GUIDE: CHANGE

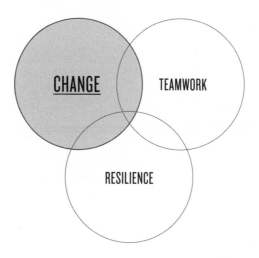

Now it's your turn! I invite you to reflect on the ideas, tools, case studies, and insights provided in the previous three chapters, all centered on the leadership concept of *change*. The following prompts will guide your reflections as you critically evaluate the relevant lessons and how they can best help you.

To target your answers and make them as immediately applicable as possible, I advise you to think about your answers only as related to your current role as a leader.

As you're reflecting, think of your own examples, case studies, and maybe even a few animal stories! Jot those down in the white spaces around these prompts. They will come in handy as you refer back to this book and pass along these lessons to others in your life.

This is how you truly personalize and implement the wisdom from the wild as it relates to change.

1. The idea that resonated with me the most from Part 1 is:

_____

_____

_____

_____

2. The biggest opportunity regarding change I can see is:

_____

_____

_____

_____

3. The biggest challenge regarding change I foresee is:

_____

_____

_____

_____

4. I am most excited about:

_____

_____

_____

_____

5. I will take action and hold myself accountable by filling in one to two (no more than that!) ideas in each box below, building off of the ideas in Part 1, to be implemented in the next three months:

STOP	KEEP DOING	START

6. I will pay it forward and help others whom I lead and influence by:

_____

_____

_____

_____

PART 2

TEAMWORK

"IF YOU WANT TO BUILD A SHIP, DON'T DRUM UP PEOPLE TOGETHER TO
COLLECT WOOD AND DON'T ASSIGN THEM TASKS AND WORK, BUT RATHER
TEACH THEM TO LONG FOR THE ENDLESS IMMENSITY OF THE SEA."

—Antoine de Saint-Exupéry, aviator, poet, author

Teamwork.

The truth about how things get done.

The behind-the-scenes reality of every effective leader.

The idea that no matter how educated, experienced, passionate, or motivated you are, you need to lead a healthy, functioning team to get the job done.

You work hard to lead people with integrity, transparency, and trustworthiness. But ultimately the success of your company's mission and the completion of quarterly goals hinge on one basic premise: how well the people who surround you work together as a team. No amount of incentives or other motivators can replace healthy, functioning teams.

I know how invested you are in your teams' success because I've heard it directly from leaders throughout the years. Leaders have told me that they are "passionate about pulling together a team for a project and having it be incredibly successful (and fun to work on!)." Others have shared with me their passions for "how teams can find it in themselves to go above and beyond" and "inspiring teams and people to follow their own passions, to dream big, to lift up their eyes to see all of the possibilities and strive for excellence."

And to accompany these strategic, visionary, laudable aspirations, how does this occur in reality, in the day-to-day, when the boots are on the ground and the work must get done? Leaders tell me that they know "teamwork will be successful if participants commit to being fully present and contributing individual experiences and perspectives."

If everything else is stripped away, organizations are simply groups of people trying to work together as best they can toward a common goal. It doesn't matter if it's a multinational corporation or a small, local nonprofit—people working together effectively in teams is a variable independent of the organization's tax status. Companies spend millions of dollars on technical systems and equipment, but if their people can't work together effectively as a team, they might as well close up shop.

The same is true in nature—animals do not survive and thrive on their own. While it varies among species, habitats, and ecosystems, multilayered teamwork can usually be found happening for everything from building nests and protecting the young to finding food and avoiding predators. Animals may not seem to employ our deliberate tactics to build and nurture teams, but they rely on effective teamwork all the same. And just as people are not all built alike, animals also bring different aptitudes and abilities to the table, and together make up a healthy, functioning ecosystem.

I've yet to meet two cats, two dogs, two meerkats, two frogs, two hummingbirds, or two humans who behave exactly the same. Our differences are what make us unique, and those combined differences make us a stronger team. A successful leader can recognize, seek out, and champion these unique perspectives and identities to forge healthy, functioning teams. When this intentional effort is made, the leader knows that the outcomes and impact of a team will not only grow, but grow exponentially.

Because teams built from individuals whose leaders see them for who they truly are, and who they can be together, are a force of nature.

Behind the Scenes with KiwiE

For the opening of a large expansion at the zoo, we invited the press, colleagues, and guests of honor for a VIP event that featured trainers with different animals mingling in the crowd to get everyone excited about the new exhibit. At eight months pregnant, I was mingling, greeting, and celebrating, all while navigating the crowd with a rather large belly.

In one conversation, I found myself happily chatting with guests and standing next to a trainer holding a young gibbon. I was beyond thrilled that our new exhibit would feature gibbons, which are small apes with long arms and a loud, melodic, musical call that can be heard throughout the zoo. They like to spend their time swinging from tree branches or extended ropes high above your head, but for now, this young gibbon was sitting in her trainer's arms.

I was laser-focused on explaining the expansion to our important guests of honor, so I didn't even notice the sideways glance the gibbon must have given me. But suddenly I felt two long, furry arms lightly clasp around my neck, was pulled briefly to the right, and felt a slight jolt as a small furry body landed on my pregnant belly. I looked down to find a gibbon staring up at me before she resettled, leaving her arms around my neck, and sat contentedly on the "shelf" of my extended shape.

To this day, I tell my daughter that being sat upon in utero by a gibbon is probably why she has always wanted to be a zoo veterinarian.

THERE IS NO ONE-SIZE-FITS-ALL APPROACH TO TEAMWORK

(Right, Naked Mole Rats?)

I am truly a scientist at heart. No matter how my career progresses and my work evolves over time, science will always be my jam. And the more unusual, gross, terrifying, or downright strange, the better.

This innate scientific curiosity may have also been known to completely take over and momentarily distract me from all other responsibilities over the years.

Not long ago, as I was heading into a facilitated workshop with clients in Texas, my focus was suddenly drawn to a small animal crawling across the sidewalk. My first scorpion sighting in the wild! I stopped in mid-stride, grabbed my phone, and knelt down to find the best picture angle and study this creature on its own terms. I barely managed to tear myself away to get to the meeting on time.

Many years before that, while driving in a caravan through New South Wales, Australia, I didn't even realize I'd yelled "*STOP!*" until I heard the cars in the caravan screeching to a halt and found myself sprinting across a field, camera dangling from my neck, to capture a picture of a goanna—the largest lizard I had ever seen in the wild. The goanna was climbing up a tree trunk and was almost as big as me. I don't know how long I was standing beside that tree, but I was eventually jolted back into reality by a few yells from my colleagues to come back to the car.

One September, some colleagues and I were walking across a bridge to visit a castle outside of Copenhagen, Denmark, when we suddenly flocked to one side of the bridge, peering into the waters below. Never mind the historical significance of the man-made structure in front of us. Our eyes were riveted to a large algal bloom that hadn't been seen for years in that area. Totally enamored by the huge mass of single-celled green organisms floating in the water, I missed the beginning of the castle tour.

I'm especially drawn to wildlife that is so unusual it defies even our imagination . . . like the animal first described by a German naturalist in the nineteenth century who "assumed from its unprepossessing appearance, the sagging nude flesh, the teeth growing straight through the skin, that it was a diseased or mutated individual of another species."[1] More than a century would pass before we began to learn more about this animal, and now it has captured even the interest of researchers studying cancer, after they realized these tiny animals are immune to the disease.[2]

When I first started noticing this unusual, tunnel-dwelling critter in zoos around the country, I became hooked on finding them whenever I could. I loved watching them either roam their tunnel system or lie in a pile of bodies, all stacked on top of one another. When I asked friends who have recently visited a zoo about which animals they like seeing best, I hear the expected answers: lions, tigers, bears, monkeys, rhinos, zebras, elephants, giraffes—all the big, charismatic megafauna that a zoo is typically known for. Probe a little deeper, and I may hear about alligators, turtles, koalas, emus, African painted dogs, manatees, parrots, penguins—all fascinating in their own right.

"But did you see the naked mole rats???" I'll exclaim.

To which the typical response is: "The *what?*"

Every Team Is as Unique as Its Members

Understanding the concept of teamwork starts when we're young. Chances are, you've either participated on a team or supported one from the sidelines. Perhaps it was a big, competitive team operating on a national level, or maybe it was a local, community-based, volunteer-run team existing solely for fun and camaraderie. It could be a sports team, an academic team, a speech and debate team, or a team with any other focus that could unite a group of people toward a common goal.

The experience we gain from being involved on a team is unparalleled. Teams are where we can learn about what we're good at (and what we're not), how to practice and receive feedback, when to step up and take control, and how to step back and let others shine. A team setting may be our opportunity to take instruction from someone we don't know, own up to mistakes that affect more than just ourselves, or truly operate outside of our comfort zone. And a team may be the first and best chance of learning how to work with others who might be exactly the same or far away on the other side of the scale from how you operate.

Being part of a team is almost inevitable in any workplace or organizational setting. Right after you land a job, go through onboarding, and settle into your role, the next rite of passage, I have found, is to join a team within the organization. Later, when you assume a leadership role, go through a leadership development training program, and settle into your role as a leader, the next rite of passage is to *lead* a team. And sometimes you get the chance to lead a team well before you were designated an official leader by your company or had any sort of formal training, and as a leader, you will still be an active participant on teams.

At times, team experience comes along with an established set of expectations, well-defined roles and responsibilities, and a clear path to outcomes. And at other times, team experience comes down to just rolling with the punches.

All of these things are common experiences on a team. Yet, regardless of where you fall on the spectrum of team leadership, participation, and overall experience, I guarantee your next team will be different from the last. And each experience with a new team offers a chance to dig in even more to the concept of teamwork.

As a staff member at the aquarium, I knew they counted teamwork as an important part of the company's culture, and I was extremely vocal about wanting to take advantage of every opportunity to learn and grow. I was constantly brainstorming ideas, researching projects to get involved in, requesting to present at and attend conferences, seeking permission to participate in new learning opportunities, and of course, keeping a lookout for any team that was recruiting new members. Finally, a spot on the aquarium-wide product development team opened up and I was given the chance to represent our department.

Beyond my assumption that it involved something about "products being developed" for the aquarium, I had no idea what the product development team's function was. But I knew I would be serving on a team with representatives from every other department, including fishes, marketing, marine mammals, and guest services. New people, new ideas, new challenges, a chance to have a new impact! I was very happy.

I gathered my notepad, favorite pen, and water bottle, and headed off to my first meeting. After taking a few shortcuts and a few wrong turns in an unfamiliar section of the aquarium, I finally found my way to the correct meeting room, only to find nearly all the seats taken. As I quietly squeezed into my seat, the team members continued chattering among themselves until the meeting was called to order. I looked around the room and saw only a few familiar faces.

The meeting concluded exactly one hour later, and we were all dismissed. I looked down at my notepad, where I had scribbled only a few sentences, and was at a loss. I had not said a word during that meeting. It was far more intimidating than I had anticipated, full of people who had many more years of work experience and shared longtime personal relationships. I didn't understand the flow of the meeting or even the point of the team.

I was at a loss for how to contribute. I felt totally inadequate and frustrated. I was terrified of letting my entire department down. I felt defeated.

The next week I headed off to the meeting again: same time, same place, same result. I really wanted to contribute—to say something helpful, something insightful, something meaningful. Really, just to say something. Anything.

Instead, I just sat there. I scribbled a few notes in case I was called upon or needed to look like I was participating, but nothing came out of my mouth.

And so it went the following week. And the next. And the next.

After about two months of frustration and private tears, I found myself sitting across my manager's desk in her office, explaining my feeling of total failure as a team member. Nodding in a supportive but firm way, she asked me to tell her more.

I explained that I felt too nervous to ask the team leader for insight, afraid of how that might reflect poorly on me or my department. I went on about how I was too proud to admit to the team leader that I didn't know how to contribute meaningfully, or really at all. I was overwhelmed with the reality of this team—it didn't look anything like the teams I had studied in college, didn't behave anything like the work teams I had been a part of before, didn't follow any of the rules or procedures I thought teams were supposed to follow, and didn't present any familiar signs of what I thought a team was supposed to be.

Where were the formal onboarding procedures? Wouldn't someone tell me how to participate? How could I get to know the other team members without a team-building experience? How was I supposed to learn all the team rules? I had been resolutely looking for road signs that would tell me how I could participate, what our goals were, and what kind of input I was supposed to be providing. But the signals I received were foreign and frustrating.

My manager patiently listened as I vented all of my feelings, without judgment or pushing back. She didn't say a word as I confessed my biggest fears: that maybe I wasn't cut out for this team. Maybe I simply wasn't going to be successful in this situation, and it was best to step aside so someone else could fill this vital role. I was unable to understand and execute on my role within this team.

The words came fast and furious, spilling out of my mouth. Then I sat expectantly, waiting for my manager to tell me that I was right, that I had stepped too far—that it was time to withdraw from the team.

Instead, she was quiet and let the emotions in the room subside. Then as she began to speak, the gentle acceptance I expected was replaced with a strong, resolute tone.

"Julie, I understand you are frustrated," she said. "And I can hear the defeat in your voice. But I chose you for this team for a reason. You are the right person to represent our department on this team and that's what you will do."

"But I don't know how to . . . I think someone else would—"

Her eyes held mine, and something in her look told me to stop talking. "You are not quitting this team," she continued. "Period. Together we will come up with a plan on how you can contribute. We will figure out a way for you to feel like and really become a meaningful member on this team. But I am not replacing you. And that is final."

There was something about the way she said those last words that snapped me back into reality. Once I didn't have the option to quit, I resolved to figure out a way forward. She was exactly the leader I needed her to be in that moment—both supportive and fully capable of holding me accountable. She also knew what the team needed, both from me and from our department. She never shared that directly with me—was it my perspective as a young professional? As a relatively new staff member? My thoughts as a woman? The respect I held from our other department members? How passionate I was about our mission? Instead of leading me directly, she allowed me to craft my own direction by taking away my excuses and created a fire in me to participate in my unique way as an active team member.

Together we set my goal for next week: Say something. *Anything.* Not necessarily pithy or groundbreaking. Just say anything to break through the fear built around me like a wall, blocking my ability to participate fully on this team.

I don't remember the topic of the next meeting, but I do remember that my interactions with the rest of the team were more casual and collegial. Once I had decided that I was going to show up, use my unique voice, and be an active member, my team members didn't seem as intimidating anymore. And we began to engage in collaborative discussions, learning more about one another, and how we could accomplish our team's goals.

After the meeting, I headed directly back to my manager's office with a big smile and sat down confidently in the chair.

"Well?" she asked with an expectant smile, already sensing that it had gone well.

"I did it!" I said with a self-assured smile, sitting up even taller.

"What did you say?" she asked.

"I have no idea. I can't remember," I said. "But it was something!"

As I continue to learn over and over again, with each team comes new opportunities and challenges. No two teams are alike, and no two team members are alike. Every team creates, operates, and produces results in its own way, and every team member has an integral role that only he or she can fill. But I have always carried with me that experience of being the team member who feels lost in the woods, and I keep a lookout for anyone who might be feeling the same.

Which brings us back to naked mole rats.

These animals may not be the ones first chosen for the cuddly animal marketing campaign, but to me, they are amazing. I stand and watch them for a long, long time—well past the time when a typical person might get tired and move on to the next zoo exhibit. They're my go-to animal in workshops when I do a get-to-know-you activity called Two Truths and a Lie. "Naked mole rats are one of my favorite animals," I say, and people usually pick that as the lie. Either they've never heard of a naked mole rat, or they can't possibly believe this non-cuddly critter could be someone's favorite.

My daughter had a naked mole rat stuffed animal (named "Moley") by the time she was three. My son, at age four, was gifted the book *Naked Mole Rat Gets Dressed* by Mo Willems. I might also have one or two naked mole rat stuffed animals hanging about my office . . .

My big, cuddly, furry, sleeping cat on my office chair doesn't even realize he's competing for my attention with a stuffed animal close by that represents a mammal with tiny eyes, a lack of hair, big teeth, and pink wrinkly skin. I'd never say that out loud because they are each unique in their own way and I love them both.

But he kind of is.

Evaluate Your Effectiveness as a Leader

Teams are multifaceted and multidimensional, which means they're different everywhere you look. Some teams are formed for specific reasons and exist for certain periods of time. They are bound by specific key performance indicators (KPIs), and their progress is measured closely until they're disbanded, having accomplished their purpose.

Some teams are part of a company's operational design, so they exist on an ongoing basis. Their purpose could be directly tied to the organization's operational or financial excellence, and they were created to execute tasks that otherwise would go undone.

Some teams are strategic in nature and exist to brainstorm what could be—what the future could look like and the innovation necessary to get there.

Some teams comprise people from both within and outside the company, including community members, funding partners, or key customers that can positively influence company initiatives and help avoid wasted time and money.

Some teams are highly functional, and others are not. This may be the result of team members failing to adhere to or implement the team's intended purpose, or it may be a result of ineffective leadership.

All teams have some things in common, however: They exist for a specified purpose, they include a defined number of people, and they have a leader. The leader may not always be the most vocal, the most dominant, or the most outgoing. But the leader is a critical component to the effective functioning of a team.

Whether you have led countless teams in the past or are stepping up to lead your first team, I want you to consider two things: First, as explained earlier in this chapter, every team is unique and thus requires an individualized approach, involvement, and dedication. And second, to provide your team with what it needs to succeed, you must begin by benchmarking your team leadership skills.

Naked Mole Rat SWOT Analysis

If you've been in the game for a while or done any strategic planning at all, chances are that you've encountered the SWOT analysis, an effective strategic thinking tool that leaders undoubtedly encounter many times throughout their career. A SWOT analysis has a number of different applications, but here we will use it to benchmark your personal leadership skills as they relate specifically to leading teams. And since we've been discussing naked mole rats as an animal example about how team members—and their leaders—are unique (and should be!) we'll call this the Naked Mole Rat SWOT Analysis. That way you can truly differentiate your responses here from those you would typically give during a board meeting or the annual strategic planning retreat.

As I advised in the section on change during the first step of the Mangrove Method, it's imperative for leaders to check in with a personal assessment of their own team leadership skills. You can't know where to focus, how to be truly effective, or even who to put on your team if you don't stop first and evaluate your team leadership skills. And if you can do this thinking about your leadership of a *specific* team (or team to be), even better. Every team can benefit from its leader completing such an assessment.

A SWOT analysis provides a framework through which to objectively and succinctly breakdown your strengths (S), weaknesses (W), opportunities (O), and threats (T). Remember, this is a personal assessment, so you can keep these results to yourself or share them as broadly as you see fit. It could be incredibly empowering for your team to engage in a transparent, honest discussion about your SWOT analysis. Team members could also benchmark their own skills as related to their participation on the team. However, sharing your analysis with others is not necessary for it to be effective. Sometimes you can be even more honest and up front with yourself when you will be the only one ever to see the results.

STRENGTHS (INTERNAL)	WEAKNESSES (INTERNAL)
OPPORTUNITIES (EXTERNAL)	THREATS (EXTERNAL)

Naked Mole Rat SWOT Analysis

To begin a SWOT analysis, first narrow your thoughts and responses in scope. For our purposes, you are going to be thinking about your team leadership strengths, weaknesses, opportunities, and threats as they relate to a specific team that already exists, or a team that you are about to lead, or a team that has yet to be formed of which you will be the leader. Thus, a Naked Mole Rat (team leadership) focus.

Next, you should jot down ideas as they come to mind, without overthinking. You don't yet need to be concerned about which ideas go where—just write them down as they come to mind. I am a big fan of sticky notes for this activity; the variety of colors can help you categorize your thoughts, and the small size limits you to one idea per sticky note. Aim to write down fifteen to thirty ideas in total. Generating fifty ideas may seem like a good idea now, but it could dilute the targeted nature of this analysis. Plus, it's hard to implement new ideas when you're confronted with myriad possibilities for both things you do well and ideas on how you could get better.

This process activates multiple learning areas of your brain and helps you focus on a measurable analysis of your ability to lead this particular team. Your goal is targeted transparency, so aim for between two and five honest ideas per square in the SWOT chart. If you are filling out this analysis with a specific team in mind, reflect on your current leadership skills as they relate to that team. If you are filling out this analysis regarding your team leadership skills more generally, be sure to write down only ideas related to leading teams. Beware the distraction to evaluate your leadership skills in general—that, too, will dilute the power of this analysis.

The top two boxes in the chart focus on your own *internal* strengths and weaknesses. If you already know the team's purpose, ground rules, team members, etc., you can be very specific here. But even if you don't know all these components yet, you will still be able to generate meaningful ideas that accurately capture your current strengths related to team leadership. Here are a few guiding questions to help you generate thoughts:

STRENGTHS

- What are my unique skills as a leader of teams?

- What strength in particular will be an asset to this team?

- What positive feedback have I heard in the past regarding my team leadership skills?

WEAKNESSES

- What do I struggle with as a leader of teams?

- What specific weakness do I need to address in order for this team to be effective?

- What critical feedback have I heard in the past regarding my team leadership skills?

The bottom two squares focus on the *external* opportunities and threats that may present themselves related to your leadership of this team. Here are a few guiding questions to help you:

OPPORTUNITIES

- What unique opportunities do I bring to teams as a leader?

- What specific opportunity do I see for this team?

- What positive effect will occur if this team seizes on this opportunity?

THREATS

- Is there a threat from the outside that I need to be aware of when I lead teams?

- What specific threat does this team currently face from the outside?

- What will happen if this external threat actually comes to be while I am the leader of this team?

Now that each sticky note has one (and only one!) idea, begin sticking them in the appropriate boxes. You'll begin to notice a pattern: Do you have many more strengths than weaknesses? Or perhaps inside the boxes you can group the sticky notes into broader ideas. For example, do five sticky notes all have to do with the idea of navigating relationships with other departments to lead this team effectively? Or maybe with juggling the realities and expectations of your "regular job" in conjunction with the team you are leading?

Sometimes it feels like you are splitting hairs when doing a SWOT analysis. Arguments can be made to include one thought in more than one category, or it may seem like you are just putting the inverse of one thought into another category. Try not to get stuck in overanalyzing. It's not a perfect science, and it doesn't have to be.

Maximizing Your Approach to Teamwork

Benchmarking your strengths, weaknesses, opportunities, and threats as they specifically relate to and impact your team will help you understand how to be a more effective team leader. This, in turn, will ensure each person on your team can maximize their strengths, which will complement one another and help build a network of possibilities to help the team thrive.

As you write down specific thoughts and observations that can be quantified, rather than relying on qualitative thoughts that are hard to articulate or measure progress against, you create a solid foundation. You can't lead a team effectively by thinking, *Well, I'm pretty sure it's going well*; that's an anecdotal strategy and open to interpretation. Instead, you need to lead by objectively and honestly assessing what you're good at, what you're not, and where the external pressures may come from, so the team benefits in the end, functioning smoothly and achieving its desired outcomes and purpose.

Remember, there is no one-size-fits-all approach to teamwork—not in any sense of the word. Every team is different, every teammate is different, and every leader of teams is different. You were chosen as the leader of this particular team for a reason, so make sure to maximize your unique approach to teamwork.

In fact, lean in to what makes you a unique team leader. Is it your approach to meetings? Your ability to navigate potential conflict? Your willingness to create a welcoming environment? Your adherence to team etiquette? Your fun ideas and novel team-building games? Even if it looks different from the outside or it's not how someone else would lead the team, stand tall in your unique style of team leadership.

Let's think again about naked mole rats. From the outside, they may seem simple, like they're not up to much, or even a little odd. They barely look like mammals, and they certainly earn the descriptor "naked." But in reality, they are part of an enormously complex living system that allows them to not just survive, but thrive in the African heat, buried underneath the ground. Very few mammals engage in the type of intricate teamwork that the naked mole rats do.

You might not know it, but you could be leading a team of naked mole rats: a bit different, perhaps misunderstood, but certainly highly functioning in well-defined roles, committed to the team and its purpose. From the outside, people may not be able to see or appreciate the complex and complementary interactions that transpire inside your team—just as most zoo visitors have no idea that naked mole rats have an amazingly complex team interaction that works perfectly for them.

Embrace Your Team's Unique Culture

There is much to be learned from—and appreciated about—teams that march to the beat of their own drummer. Every team is a patchwork quilt of skills and perspectives, which means it may come across as confusing and could be misinterpreted by others. But as with my favorite hairless East African mammals, your unique brand of teamwork may include coordinated efforts, synergy, and productivity hiding just beneath the surface.

Naked mole rats live in colonies of between twenty and three hundred animals in an underground tunnel system.[3] They work together to build these tunnels that connect nesting chambers, food sources, and even toilet areas, and that extend up to 2.5 miles in total length. And they live together in a social structure much like bees do, with a queen, workers, and an efficient division of labor.

But their dedication to teamwork doesn't stop there. Although they are mammals, naked mole rats need each other to regulate their body temperature. While they might appear to be lying around on the job, piling on top of each other is actually a survival strategy that helps each animal stay warm.[4]

Is your team underestimated by others because of how you interact, which may be different from how other teams interact? Are your meetings an unusual standout within the overall company culture? Do you manage to come to consensus even without giant pieces of notepad paper or a river of sticky notes? Does your team simply not look like a "typical" team?

Or perhaps you're a team leader who doesn't present like a "typical" leader. Perhaps you find yourself defending your team members and how they contribute or trying to explain your unique ways of leading. Maybe people are even shocked at how effective your team is or how you managed to assemble it, as they can't fully appreciate each team member's talents and perspectives.

Don't listen to the naysayers! The positive impact of your team may not be well articulated, but the negative impact will certainly be noticed if your team ceases to function in a healthy way—or completely disappears.

Naked mole rats don't try to be anything other than what they are. They simply own their niche. They do what they do, how they do it, and they don't apologize for it. And they are extremely successful at thriving— and working together—in their own unique way.

My daughter is an avid teammate, too, a member of a highly competitive team that truly enjoys working together, united by their shared love of their sport. The team members range from twelve to eighteen years old, spanning a tender coming-of-age period that is buoyed by shared experiences among people going through the same phase of life.

Coached by a national champion, my daughter's team draws kids from a two-hour radius each year during tryouts. On a weekly basis, they practice both their sport and the conditioning drills meant to keep them in shape and injury-free. They travel for competitions, and she wears her team jacket with pride. Since qualifying to join the most advanced team in our community, I have seen her rise to the challenge of operating outside of her comfort zone and leveling up her skills to match the teammates around her. If she continues on this path, she could have the option of trying out for a varsity-level college team and even winning a scholarship to her chosen university. This team has become a part of my daughter's identity and has truly given to her as much as she has given back.

By now you have in your mind a picture of what this team could possibly be—a picture of a team similar to one that you've experienced in your life or your child's life, or even a team that you see children participating in on TV. Maybe it's a team you would like to join in the near future.

What team are you envisioning?

I bet the picture in your head is of anything but a nationally competitive, intricately synchronized team of ice skaters.

Did you even know synchronized skating is a sport?

Most people are unaware of all the inner workings and shared responsibility that has to happen for fifteen teenagers to skate at top speed to music on an ice rink while fluidly performing a routine, transitioning seamlessly between various formations, individual elements, and shoulder holds. But I guarantee that if any skater misses a shoulder hold, arrives in position a split second too late, or slips on the ice, that skater can take down the entire team. Literally. I've seen it happen too many times to count.

I'm amazed when I watch this team perform with wordless communication that can only be developed over hundreds of hours rehearsing together. When I see one girl double down on her grip as another girl stumbles just slightly—in order to keep her from falling and thus keep the whole team together—I know their teamwork is one of a kind. But that may not be apparent to the untrained eye. If you saw my daughter's team in action, they might not look like they are demonstrating teamwork, from the outside, to you. They might look like a team of naked mole rats.

My daughter has heard those comments, from kids at school who don't understand how synchronized skating works—who don't even believe that ice skating is a competitive sport. My advice to my daughter is the same I give to my clients: Keep your head held high and keep moving forward.

Only give credence to the naysayers if you've asked for their input. Otherwise, their opinion is irrelevant. They don't get a say on whether or not your team is "typical." They don't get to judge if it's highly functioning or not. You'll reach out if you want their perspective.

And in the meantime, keep doing what you're doing. Concentrate on being the best team that you can be, laser-focused on being productive, supporting one another, and working toward shared outcomes.

I was working with a client to help his team increase its productivity and outcomes. Company management felt that the team had begun to lag in productivity lately, and the rest of the organization did not seem to understand the team's role, function, or outputs. To counteract this perspective, the team leader began seizing more and more opportunities, generating more ideas, and overwhelming the team with new approaches, tactics, and projects to accomplish. Team members weren't sure what to focus on first, or how to measure and communicate the progress made.

As a result, the team was struggling to keep up, innovate, and even keep its members healthy. People were calling off work and approaching a nervous breakdown. Well aware of this situation and deeply empathetic in nature, the leader was at his wit's end.

We decided to strategize regarding the team's interactions and then communicate to the company at large about how the team was working, its KPIs, and how its results directly affected the company's bottom line. This, in turn, would positively impact the other executives' perspectives on this hardworking team, and they would better understand how to leverage the team's talents in the future. Most important, the team leader would know how to assign the right team members to balance their innate, creativity-focused leadership skills with a detailed-oriented approach to the work—which would help the team members channel their energy and establish time-bound goals to be accomplished.

It didn't matter whether the team leader led differently from the rest of the company. His style might not work for everyone—and it didn't have to. What mattered was that the team knew how to function effectively, in defined parameters, working toward the same goals, and that these efforts were clearly communicated to the rest of the senior leadership. That way, achievements could be celebrated, and progress was made toward the company goals.

Another client's team was completely bogged down with how they were "supposed to" do things. They were so busy trying to assign roles, apply appropriate corporate jargon, and establish timelines that the team members and their leader alike had completely lost sight of their original purpose. They were so busy trying to become a "typical" team that they had lost sight of why the team had been established in the first place.

Stripping away all the artificial constraints and reestablishing the team's purpose and priorities immediately freed this group to become a functioning team with an effective leader again. Then they could focus on team-building, getting to know one another and how each member could contribute, before reincorporating the official company way of doing things so they could fit within the culture and be held accountable. Once they stopped trying so hard to be a team and just were themselves, they could embrace their individuality and their unique approach to positively impacting the bottom line.

No longer trying to force rules and operational considerations on the team, and instead letting them evolve and participate fully, the leader was unleashed to manage a team at maximum effectiveness. Being aware of and championing the individuality of team members—and envisioning how they can work together in unique ways to accomplish stated goals—is the name of the game, even if it looks different on the outside (or even the inside) from how people expect it to.

Embrace the uniqueness of your team leadership journey. Don't shy away from your naked mole rat moments. There is no one-size-fits-all approach to teamwork, and there doesn't have to be. Teams are as unique as their leaders, their members, and their purpose.

The world—and your organization—needs naked mole rats.

Unbreakable Law #4 Pro Tips

➤ Teams are as unique and distinctive as the leaders and people (or naked mole rats) who form the team, each fulfilling an individualized and integral role.

➤ Understand the impact of your strengths, weaknesses, opportunities, and threats as a leader on each team you lead, to help team members maximize their strengths and work together to help the team thrive.

➤ Teamwork can be misinterpreted at first glance, when just under the surface are miles and miles of coordinated efforts and productivity that result in a healthy, thriving team.

Behind the Scenes with KiwiE

Dolphins are conscious breathers, which means they quite literally have to think or decide when to breathe.[5] So, to sleep, they shut down one half of their brain while the other half stays awake and aware of the environment around them.

The dolphins were some of my favorite animals to visit during late nights at the aquarium. When I needed a break, I would head to the underwater viewing area, and suddenly multiple dolphin faces would appear at the window out of the darkness. We would look at one another for a moment, and then I would turn to my left and start jogging along the windows. They would follow as we began our impromptu game of tag. Abruptly, I would turn around and head the opposite direction and watch as they followed along. I'm quite sure I ran out of energy far before they did during our underwater races.

SURROUND YOURSELF WITH PEOPLE WHO ARE NOT LIKE YOU

(Termite, Meet Giraffe—Giraffe, Meet Termite)

Hanging on my office wall is a simple black-and-white line drawing of a picturesque moment on an African savanna, with birds flying overhead, giraffes striding across the grass, lounging lionesses gazing into the distance, and termite mounds rising from the dusty ground. Because this drawing is tucked among more vibrant pictures and paintings, you might not even notice it. It sits in a large rectangular frame, which long ago had a fitted glass covering and actual attachments for hanging on the wall. It has lost most of these components over time and is currently jury-rigged into position. Although it is not the fanciest picture and certainly shows its age, nothing has displaced it just yet. Because no other "thing" in my office so completely and succinctly captures my professional evolution as this picture.

This image was originally designed to be colored in, by children or families wanting to learn more about how each group of animals on the African savanna are uniquely designed to fill their own unique niche. It was an integral component of the first big important project I ever worked on for my job at the zoo, when I was assigned to help coordinate an annual event to celebrate wildlife and wild places. This event invited weekend visitors to experience fun and games while perusing booths set up all throughout the zoo's grounds by local corporate and community partners.

Feeling enormously grateful for and excited about this responsibility, I started by generating ideas, building off of the zoo's past events, and creating lists and action steps. Immediately I could see just how many volunteers (at least thirty), departments (five), animals (twenty), and organizations (twenty-five) it would take to pull this off. My first big important project was going to require extensive teamwork on a level I had yet to manage.

Although I'd attended events like this in the past, it's an entirely different experience when you are partly responsible for running one. Being accountable for not only the operation of this event, but also the experience of the guests and community partners, began to weigh on me. I would need to develop and communicate a schedule, invite the local partners, make sure they had parking passes and knew when to arrive, coordinate lunch breaks for everyone, and plenty more. Plus, I had to think of something we could produce and give away to all the attendees. And this was well before the era of text messages and pervasive email!

The deeper I got into planning and thinking about how I was going to pull this all off, the more I was able to envision a giveaway idea: How about an informational coloring page capturing the essence of an animal ecosystem? After all, for this event to happen, everyone would need to work together using their unique skills in their specific area of expertise— just like the animals surviving on the African savanna, each inhabiting its unique niche.

An ecosystem would never work if all the animals had the same skills and needs, just as society would never work if people were all good at the same thing. As is the case with wildlife and wild places, to accomplish the work we humans do, everyone needs to find their niche and do their thing. That's how ecosystems stay healthy, how work gets done, and how teams thrive.

Is it (usually) harder to work together with people who are different from you? Yep.

Do lions always get along with hyenas? Nope.

But is the ecosystem healthier and more productive when a variety of species do work together? Yes indeed.

That's how nature is designed to work.

Would you want to question whether a vulture knows what it's doing?

Or would you prefer eating the dead animal this scavenger has found and help keep the savanna free of decaying carcasses that could harbor diseases and harmful bacteria?

Me neither.

Our Differences Are Where the Magic Happens

There is just something comforting about people who "get" you, who understand what you are thinking without you having to explain it all to them. They can finish your sentence without batting an eye. They can stand in for you when you can't be there, because you're confident they would make the same decision that you would.

And then there are the people who are different from you, who understand life through a completely different lens. They don't always take you at your word, but instead want to know how you formed your opinion or where you got your facts. They challenge you on the decisions you make before agreeing—or apathetically going along with it. They communicate in a different way, with more (or fewer) words. Perhaps they need to take notes on everything with detailed precision, while you remember it all in your head—or vice versa. They are driven by different motivations and may even hold values that are different from yours.

But this is where the magic happens. Our many differences are what nature has intended. Take a look outside. Are all the trees the same?

What about the birds? Are the ants that scurry by seeking the same thing as the worms slithering past the other way? If all species of snakes were eating the same local population of rabbits, how long would those snakes survive? If hawks and owls were both hunting for mice during the day, how successful would each species be?

Animals, including humans, survive and thrive together because we are *not* the same. Diversification is a biological necessity for organisms to survive and thrive in the world. And this provides the possibilities for collaboration and teamwork—whether intentional or driven by instinct—to generate a successful thriving ecosystem across both wild and business landscapes.

I evolved originally to see the world through a very black-and-white, data-driven lens. My science coursework taught me to trust research and take copious notes. I often will add something to my to-do list *after* I've completed it, just so I can have the satisfaction and visual confirmation of crossing it off. Yet, along the way, I've discovered (and leveraged) an impulsiveness and ease with some level of uncertainty; I've learned to become just as comfortable trusting my gut as I am basing a decision on reported facts. I've embraced the value of questions—not only about what the numbers say, but also about how people interpret them. I want the research, but I am no longer constrained by it. I feel comfortable in being uncomfortable, because I have learned that what follows is bound to lead to something new and exciting.

Fostering my love of the gray areas of opportunity and change has helped me enormously in my career, not only in achieving my intended goals, but also in learning how to work with others who are different from me. I can now flex back and forth along this sliding scale to complement their various approaches—adapting a "show me the numbers" approach when I'm surrounded by people who want to go with the flow, and being the champion for gut instinct when I'm surrounded by people who are buried deep in the numbers and have problems moving forward with a decision.

How did I *really* learn this? By being a part of teams with people who were different from me.

For one project at the aquarium, our close-knit team was extremely close to launch after working together for six months—so long that we all understood how to communicate with one another, anticipate other team members' input, and trust our collective judgment. Then a person was added to our team who brought very high-level, strategic, almost eso-teric questions into our midst. We would be talking through a concrete, boots-on-the-ground idea, about ready to agree, when inevitably he would ask, "Sooooooo . . . what is the point of all of this? Why are we going down this pathway? What does this *really mean*—not just for our bottom line, but for our mission? Is this really how we want to represent our brand?"

The collective groan would be almost audible. Even if he couldn't hear it, I'm sure he could read the frustrated expressions on our faces. But he wouldn't rescind his questions. He was totally comfortable with our angst and believed deeply that his input would make our project better.

Around the room, the rest of us would exchange annoyed glances and a few eye rolls before doubling down on why this concrete idea was the best way to move forward. We would argue how it built on past efforts and share the quantitative evidence we had used to chart this new path.

He never gave in. And neither did we. But together we forged an idea that was an amalgamation of our combined input. When we finally imple-mented the idea, we had to admit that his input had made it better.

This individual is now a CEO and someone I seek out for trusted input because I know he'll give me thoughtful, thought-provoking ideas and always tell it to me straight. He is incapable of being anyone other than who he is, and he does not apologize for it. Instead, he has forged a career of positive impact and has maximized his influence and visionary thinking the best way he knows how—by surrounding himself with concrete think-ers like me to keep him balanced.

As a consultant now, I can play a role very similar to his role on our aquarium team. Many times I am brought on board for a project specif-ically *because* I think differently: I come from outside the company and am not bound by internal stories that might derail potential ideas. I bring experience from a variety of industries, and I can poke holes in ideas with the overall intent of moving the project forward in both practical and innovative ways.

Working with the CEO, the board of directors, and senior leadership, I can facilitate visionary conversations with tangible outcomes. I help ensure their ideas are fully brainstormed and then relate them to real possibilities for the organization.

Working alongside the staff of all levels, I can harness input, encourage equitable participation, and build trust in both the process and the outcomes. I am always on the lookout for people who operate like I used to, who search for data, want more research, and need reassurance that decisions are built on the company's foundation rather than from scratch. I also recognize the big-picture thinkers, who are already four steps down the road, thinking about next steps or implementing ideas. I need them both—and everyone in between—to ensure the project reaches its highest impact and potential.

I adore standing in front of a team, explaining from the start that I know there are detailed process thinkers who will think of all of their questions a few days from now, and there are fast-paced visionary thinkers who are already strategizing though we haven't started our discussions. I tell them it's *my* job to help them all participate equally and unite them on the pathway that will ultimately lead to actionable outcomes. So, we begin by building trust and acknowledging the range of perspectives and modes of thinking in the room.

In any consulting engagement, part of my role is to make sure when I leave that the team is fully functional, so it can implement the ideas and actions identified. Without team members who bring different strengths and can carry out different tasks effectively, there is no functioning team. Without a functioning team, nothing happens—no follow-through occurs. And then the company goes back to square one or even further, eroding any progress that had been made.

The key to all this is the team members who truly complement and enhance one another. It's the multiplier effect—the idea that $1 + 1 = 3$, not just 2 with a healthy, functioning team. Strategically putting together people on a team can make you a more effective leader, with a stronger team, heading toward a brighter future.

Create a Diverse Team to Provide Opportunities for Growth

Regardless of your age, years at the company, or years spent studying how to be (or simply being) a team leader, you should approach the next team as if it's your first. And depending on your inherent nature and experience, the associated professional risk, and the purpose and critical nature of the team, this approach will either invigorate and inspire you, or bring uneasiness, angst, and apprehension.

New leaders, feeling apprehensive, often are tempted to populate their team with people with whom they share similarities and communicate easily. It may seem like a good idea to surround yourself—in the office, at lunch, or at the company picnic—with people you agree with to ensure less friction or potential divisiveness. When the people working closely with you are on the exact same page, it's not only more comfortable but sometimes just plain easier. It feels good to make decisions quickly that everyone agrees with; when everyone thinks like you, there are fewer challenges and less pushback.

Don't misunderstand me; "new" leaders are not always "young" leaders. You might have lived seventy years or more, but this is your first time leading a team of this particular nature. You could have been at the company thirty-plus years, but this is your first opportunity to lead a team targeted on this particular goal. Or maybe it's completely the opposite: You've been leading teams for so many years that it feels like a well-oiled machine, but you just don't have the energy to put into it anymore.

Perhaps the team is so important to the company's stated goals that it seems like a huge risk to populate it with people who may bring too many challenges. Or maybe the team worked so well last time that you're hesitant to "mess with a good thing," so you just keep them together for the next project. Let me assure you right now: That's not the path to take.

Picture the African savanna. In your mind's eye, what landscape do you see? Which animals come to mind? I love how writer Sindya N. Bhanoo of the *New York Times* describes it: "The African savanna has a cornucopia of majestic creatures—lions, elephants and giraffes among them. But behind the scenes, it is the tiny termite that fuels much of this diversity, a new study reports."[1] In this research study, scientists noticed large patches of trees, plants, and green grass up to 30 feet in diameter and spaced hundreds of feet apart. When they dug a little deeper (quite literally), they discovered millions of termites living underneath these lush green areas.[2]

Among humans, termites are viewed as pests, usually found eating the dead wood used to construct parts of buildings such as our schools or homes. But in the animal kingdom, this "destruction" is precisely what is needed! When the termites consume dead wood and plant materials, they release the nutrients that are needed for new trees, plants, and grass. The large animals that come to feed on this new growth, like giraffes, add their own nutrients in the form of urine and feces that also support plant growth. Thus they (both the termites and the giraffes) are directly contributing to the creation of green spaces—and food sources for each other—in the otherwise sparse African savanna.

When envisioning the savanna, I can imagine walking in the vast expanses of wide-open grasslands, punctuated every so often by trees. I wish I had the power so many children dream about—being able to talk to animals—as I would love to introduce two animals: "Termite, meet giraffe. Giraffe, meet termite." I would ask each to thank the other for its distinctive role in creating the eye-catching green spaces conducive to plant growth, which each creature in the food chain relies on.

Termites and giraffes may not *know* that they need each other, but they do. Each plays a vital role in ensuring the other has the nutrition they need, directly impacting the food available for the other to eat. Although they exist as two starkly different animals with seemingly nothing in common, their actions help each other survive and thrive.

Of all the types of teams I've worked with, been a member of, and led, the one in my experience that relies most on differentiation for its success is the board of directors. Whether for a large for-profit corporation, a national nonprofit association, or a local, membership-based chamber of commerce, a board of directors offers challenging and intriguing options for participation and leadership. Serving on a board is often very educational in terms of learning new rules and acquiring skills for interacting with team members with a wide range of experience, expertise, backgrounds, perspectives, and interests.

My number one piece of advice to anyone looking to further their professional experience or positively impact their community is to find an organization you are passionate about, join a committee, and eventually offer to serve on the board. I was elected to my first director position on the board of a national nonprofit association at twenty-four years old, and I never could have imagined how much I would learn about both myself and every aspect of the company for which I was now responsible. The experience truly shaped my leadership style, as I learned how to work effectively with people who are *not* like me and why I should always strive to be surrounded by people with different skill sets.

When you join a board of directors, you quickly discover that not all boards are created equal. Does this board use Robert's Rules of Order?[3] Do the meetings always start and end on time? Is there an official orientation to kick off your term of service, or perhaps a certification course you must take before serving? Or does this board encourage casual interaction, with chairs pulled from various rooms at the start of the meeting, cookies passed around that someone brought to share, and a chairperson who has to be kept on task so the agenda is followed? Is it the sort of board whose meetings have been known to be interrupted by a penguin walking into the room? I've been a member of all these boards (including the one with the penguin visit).

Just as there is no one-size-fits-all approach to teamwork (see Unbreakable Law #4), boards of directors may look strange from the outside when effective, impactful leaders surround themselves with a collection of very different people. Developing that sort of team within a company hierarchy can be a long and challenging process compared with building a board of directors. Board positions are often bound by term limits, so the opportunity to bring in new people repeatedly presents itself. And depending on the articles of operation, the number of directors may be flexible, offering yet another entry point onto this kind of team.

Because a board is the public face of the company in many ways, the opportunity to reflect both its customers and the relevant community is tangible and ripe for the taking. For instance, one of my clients, a large, privately held company based in the southeastern U.S., appointed its first woman to the board of directors after more than fifty-five years in existence. Beyond valuing the particular expertise and experience she could bring to the board, the company had proactively chosen to be more reflective of the organization and its clients. Another client was adamant about representing the indigenous people who traditionally held the land where the organization sits, so began by appointing a tribal member to the board of directors. The company gained an increased understanding of—and an important, authentic relationship with—a critical stakeholder and community member.

Both of these clients took the first and necessary steps toward actively recruiting and integrating people who could bring diverse perspectives and input, thus positively launching the organization toward future growth.

Hey, Giraffe, You Need a Termite

As a leader, you have worked to form a well-functioning team. Now it's time to acknowledge (admit?) that you need to find your complement. The person who can reach the people you'll never be successful reaching. The person who can figure out an entirely different path to the team's goal. And that may just be the person who seems the most different from you at first.

In short, to thrive as a giraffe, you need a termite.

At my initial, day-long meeting of the first board on which I served, I was familiarizing myself with the financial statements and learning the unwritten rules and expected behavior. Sitting at a giant U-shaped table in a giant conference room in California, I was impressed by the sheer magnitude of it all. It was not lost on me that these forty-odd people, representing Ohio, Florida, Oregon, New York, Hawaii, and more, were making decisions that would direct the future of this national nonprofit organization.

The chair called the meeting to order and welcomed new members of the board. We all listened as directors reported out on various topics and discussions ensued. Then the discussion got heated over an issue that the board had been dealing with for a while, with some members taking copious notes and others sharing their thoughts at every turn. I was observing the multitude of subtle interactions among directors, which seemed to come out of long professional friendships or strategic alliances, as both support and dissent peppered the room.

After eight long hours of being trapped in a windowless room—working, debating, processing, decision-making—one final challenging topic stood between us and the outdoors (and, let's be honest, happy hour). The weariness was evident on everyone's faces, and the chair was doing her best to keep us on track. Voices rose and tempers flared. The exhaustion began to get the better of us all. It seemed we would either go well past the end of the stated meeting time or draw to conclusion without a decision—and after all of our hard work, neither idea was palatable.

Just as it began to feel like there was no end in sight, a table tent was flipped vertically on its end by a man at the far corner of the room who had not yet spoken aloud that day. At this agreed-upon signal, the chair called on him for his comment, and the man now held the collective attention of the room.

As the room began to quiet, he jotted a few final notes on the paper in front of him. He calmly extended his hand and returned his name placard to its horizontal position. Then he looked up with an expectant look on his face, and there was silence.

As the man began to speak, I noticed immediately the marked difference in his voice compared with the other board members' voices throughout the day. His tone was calm and his volume soft. He spoke with what seemed like a deliberately slow pace and said only what he had to say—not more than a single sentence. And in that one sentence, he succinctly articulated the hours-long controversy and offered a possible solution.

The room remained silent, with none of the usual rebuttal, whispering among members, or even furious typing on computers. Just silence.

I turned my attention back to the chair. She asked for a motion. A motion was voiced, then a second, and then we were voting. Five minutes later, the meeting was adjourned.

Amazing.

I turned my attention back to this man, in awe of what I had just witnessed. How had he analyzed, processed, and so effectively communicated everyone's thoughts, opinions, and critiques in just one sentence? And how had he done this in a way that would appease all perspectives, allow for a responsible decision, and permit us to close the meeting with pride and a sense of completion? He had condensed all the discussion of the day into a comment that was pointed not toward agreement, but toward consensus—which is how professional decisions get made and important work gets done. And he'd accomplished all this with a volume and tone of voice that completely held the room's attention by being quiet and firm rather than loud and questioning.

This is a unique skill. I could feel it and see its impact then, and the realization of it is still seared in my brain to this day. It is a skill I do not have. But I immediately sensed how it was valuable and could complement my own strengths. I knew instinctively that I, as a giraffe, was in need of him, the termite, to accomplish the things I wanted to on this board.

After working together on multiple projects over the years, this colleague and I now have mutual respect for the individual skills we bring to the table and how we can work in tandem to achieve our goals. When we are in a meeting or a retreat, I know the result will be magnified because of how he can articulate and persuade in a completely different way from me. If I get too wordy, I can glance his way and he will interject to eloquently capture what I am trying to say. If he is at a loss for words he will glance my way, and I can add more explanation to help others see his point. If I need to make an argument, I rely on his ability to frame it without alienating people from the discussion. If he needs to bring more people to the table, he relies on me to reach out and solicit their opinions.

When we are on the same team, I am a better leader and can have a more positive impact. I am the giraffe to his termite—or maybe some days it's the opposite. Regardless, I stand proud alongside my colleague, who has an entirely different skill set than mine, knowing that together we keep the organizational landscape as healthy and thriving as the African savanna.

The savanna is an ecosystem in which every animal plays a unique and necessary role. If you look closely enough, you'll see a whole host of teamwork principles: Elephants embrace communication. Meerkats show cooperation. Zebras exemplify unity. Hornbills exhibit dedication. African wild dogs demonstrate loyalty. Vultures have patience. Each contribution is as necessary as the next, and each animal is as different as the next. But together they create a healthy, thriving ecosystem.

In business, as in all aspects of life, each person must fill our own unique niche and role—to thrive in the here and now, so we can adapt to changes in the future.

Our differences are what make us successful as a team.

No one person is more important than the next.

Giraffes, you need the termites.

And termites? You guessed it. You need the giraffes.

Unbreakable Law #5 Pro Tips

➤ Recognizing that people experience life through different lenses and understanding how different skill sets create a successful ecosystem will help you accomplish far more as a leader than you could on your own.

➤ Intentionally seeking opportunities to grow as a leader of teams, such as serving on a board of directors, can be an unparalleled chance to learn to work with and embrace a diversity of people, experiences, perspectives, and expertise united toward a singular purpose.

➤ Successful leaders know they need a termite in order to thrive as a giraffe, that a seemingly inverse or paradoxical colleague may just be the person who can help you achieve the impact you are seeking to make.

Behind the Scenes with KiwiE

Before I worked in a zoo, I had never heard of a kinkajou. This rainforest animal looks like a monkey and is nicknamed the "honey bear," but it is neither a bear nor a primate. Related to raccoons, kinkajous are mammals with big eyes, dense fur, and a prehensile tail that can hold on to things. They quickly became one of my favorite animals to handle and show to visitors, whether out on zoo grounds or in our classrooms. Often, the kinkajou would curl itself into a ball in my arms and sleep while I told my wide-eyed guests all about this unique animal.

One evening I was teaching an overnight program and planned to show the group a kinkajou. As I approached its exhibit behind the scenes, I saw an unfamiliar sight: This normally tranquil, snoozing creature was now bright-eyed and bushy-tailed—literally! Standing in front of my group with the kinkajou, anticipating the expected feeling of holding a sleepy creature in my arms, I now held a highly energetic animal. As the kids watched with wide eyes and open mouths, the kinkajou was far more interested in traveling up and down my body with as much leeway as I would let him than sleeping, taking in the sights and sounds of the classroom at night.

It was one of those moments when everything you've read suddenly comes to life in front of you. While I knew *in theory* that a kinkajou was nocturnal, it wasn't until I went to handle this animal *at night* that I truly understood what that meant. "Active at night" really does mean active at night!

A TEAM WITHOUT A SOLID FOUNDATION IS REALLY JUST A GROUP

(Shark! Eel! Lobster! Coral?)

Did you know the ocean is full of sounds?

It's true. When you immerse your head underwater, you can hear all kinds of noises. You may be unable to identify them or even tell from which direction they come, which can be extremely disorienting. Under the ocean, you can feel a bit unmoored from reality as you know it.

On land, our ears help us determine where soundwaves are coming from, and the noises we hear become the familiar backdrop over the years of our life. Car horns honking, birds chirping, people laughing, hammers banging, music blaring . . . all these sounds and more fill our daily lives. And then you jump into the ocean, perhaps expecting it to sound like nothing, only to find it's filled with sounds of its own. But if you listen closely enough, you may be able to hear one sound you recognize.

Chewing.

When I've jumped into the ocean to look for manatees, I've first been shocked as chilly water fills my wetsuit, and then a bit dismayed as I peer through the dark water, so full of tannins that it's hard to see through. If I rest in stillness, though, concentrating on breathing deep through my snorkel, I can hear it: the distinct sound of teeth munching. I know then that manatees are nearby, chewing on their food of choice, sea grass.

The experience is similar when I jump off a boat to scuba dive a coral reef, excited to see the colorful fish swarming just up ahead. I feel the water's buoyancy as I descend to my chosen depth, grateful for the weights on my wetsuit that carry me deeper and avoid an early, unplanned return to the surface. As I focus on the sights ahead, curious to know what I'll discover this time, I can hear it: the distinct sound of teeth crunching. The parrotfish are nearby, chewing through the limestone skeleton of the coral to get to their food of choice, algae.[1]

All this chewing is amazing—if at first somewhat surprising—to hear underwater. Although both the manatee and the parrotfish are free to roam wherever they like, they usually choose to stick around a particular area for good reason: Their food source is there.

Of course, like the sea grass, the coral function as more than a food source. They are organisms in their own right, with their own life cycles, nutrition needs, habitat requirements, and adaptations to withstand threats. But what I think about in this moment—as I am surrounded by the sound of chewing—is their basic, fundamental role in the ecosystem. Without sea grass, the manatees start to leave. Without coral, the parrotfish must find somewhere else to graze. They feed a myriad of animals and, as such, provide a habitat for these animals in which to live.

If the sea grass beds and the reefs are not protected from damage caused by boats whose motors come too close, or from people whose flippers knock off pieces of coral, or chemicals in some sunscreens, then they will be unable to live themselves, let along provide critical habitat.

A clump of sea grass or a cluster of coral is more than a *group* of living things—they function more as a *team* that forms a critical foundation for ocean life. Even more, they form the basis for boating, fishing, scuba-diving, snorkeling, and other activities.

Without this foundation, everything starts to fall away.

Teams Thrive When They're Designed around Shared Fundamentals

In order to be truly effective, a team must have a solid foundation.

It must be built on solid rock.

Period.

What do I mean by a foundation? How do you know whether your team is built on solid rock? And if it's not, what is the impact?

To uncover the answers to these questions, close your eyes and imagine jumping with me into the ocean. Feel the saltwater envelop you and taste it on your lips. Now, in your mind, open your eyes and look around you. What do you see?

Perhaps you are acutely aware that your senses work differently underwater, causing you to second-guess things. You can no longer rely so heavily on what is familiar to you on land, but I will be your guide as we swim away from the boat.

Which part of the ocean did you jump into? Are you in shallow water, surrounded by bright and colorful fish? Or are you descending into deep water, past the reach of light? Are you watching a moray eel sneak around a corner of the reef to search for prey, or are you immersed in a dense kelp forest that reaches as far as you can see in every direction? Are you enjoying the warmth of the water against your skin as you float along, or are you covered head to toe in a wetsuit so thick that bending your arm takes extra effort?

Come with me this way and imagine you're swimming in warm water toward a coral reef . . . now look closer at the scene right in front of you, swarming with activity: parrotfish darting in and out of the reef . . . sharks circling . . . tube worms popping up . . . eels poking their heads out . . . stingrays hiding under the sand . . . lobsters crawling past sea cucumbers. Right in front of you are so many creatures, it's hard to focus your attention. Just when you notice a barracuda swimming over your head, a little rock shrimp catches your eye. A giant, silvery ball of baitfish shimmers to one side of you, and then parts suddenly as two majestic tarpon glide through the opening.

But just beyond the reef—can you see it?—nothing but sand stretching into the distance. No movement. No obvious signs of life.

Why is all the life seemingly concentrated here on the reef? What anchors so many species in one place?

It all begins with the coral.

There are over 6,000 species of coral,[2] which can be found around the world in waters ranging from shallow and tropical to cold and dark, 20,000 feet beneath the surface.[3] As invertebrates, they lack a backbone and are actually closely related to jellyfish.

Coral reefs are built by hard, or stony, species of coral that are basically upside-down jellyfish—large and bulbous body with tentacles "dangling" upward—living inside a limestone "cup." Yes, limestone—the sedimentary rock that is found on land as well as in the sea. The squishy, almost shapeless body of a coral animal spends its life in the protection of an external "skeleton" made of rock. And while no two species of coral are alike, together, they can build an expansive, important habitat that anchors a wide range of underwater life.

No two team members or leaders are exactly alike either. Yet together, they can build off of a shared purpose and establish fundamental operating norms that can anchor a common understanding and support the achievement of a wide range of corporate initiatives.

As the leader, you might assume that when a team is formed, everyone is on the same page—that because everyone is committing time and resources to this team, everyone knows and shares and can lean in to the team's purpose. You could assume that all the team members understand how the team intends to operate and are on board with the expected etiquette. You may take for granted that they know why they, and all the other team members, are present and how each individual uniquely contributes.

This is one of the myths about teams that I most frequently have to debunk.

In my experience, leaders tend to err on the side of assuming people understand much more about the purpose of a team than the members actually do. Leaders rightfully don't want to waste their team members' time or insult their intelligence by taking the time to state the team's intended purpose or describe how things will work. Many times, they would rather jump right in and start getting work done. In fact, leaders preparing for the first team meeting often share with me that they plan to kick off the meeting with something along the lines of: *Thank you all for joining this team. Because we all know why this team has been formed, how it will function, and the anticipated outcomes, let's just get started.*

But I've often heard members of those same teams say they have no idea why the team was formed, why they've been asked to join it, what the team meeting structure will be, and what the team is supposed to achieve. They lack understanding about the common foundation.

It's not that the team members haven't been paying attention or don't care; it's that their world is entirely different from yours. As a leader, you are focused on the big picture, while they are focused on this week's deliverables. You are driven by first-quarter goals, and they are driven by month-end realities. You have been involved in 30,000-foot conversations with the board of directors, senior leaders, and community stakeholders, while they are busy fielding extremely particular inquiries from customers or solving problems that arise out of daily operational realities.

Although you are living in two entirely different hemispheres, you remain united by the company's singular mission—and even more so when you identify the specific purpose for which the team was established. Because somewhere in the middle of these two hemispheres is where the synergy happens, where the team enacts the norms understood by all of its members in order to live up to its shared purpose.

As the leader, you see the change that needs to happen, the opportunity on the horizon that will best be addressed by this team. But the people who need to function together on the team to bring that opportunity to fruition—the people who will need to *react* to and *implement* new ideas brought forth by the team—are as diverse in scope as the animals we observed on our imaginary dive. And the rules and expectations about how the team will work and what success means are like the coral reef that supports and empowers that diverse bunch of individuals.

I once worked with a manufacturing client that had recently under-gone substantial budget cuts, which had also affected the workforce. The organization was now operating with a reduced staff but the same work-load, while also dealing with increased demand for productivity due to the forecast of a tight budget year. The leaders were adjusting the best they could, reallocating any resources they had while working to maintain a positive and cohesive culture—until a decision came down from the most senior level of leadership to form a team that would look at how to stream-line productivity.

This did not go over well with the mid-level leaders. Already frazzled while trying to keep the business afloat and planning days (not even weeks) into the future, they were exasperated by the senior management's edict, which seemed out of touch with reality. The senior leaders were looking toward the change they could see coming on the horizon; the mid-level leaders were focused intently on the day-to-day details.

How had they gotten there? The world had been spinning so fast for the entire company that the senior leaders had not been able to take the time to truly understand the reality of day-to-day, on-the-ground busi-ness operational pressures. And the mid-level leaders who would need to implement this change and deliver the results had not been part of any decision-making about the "nuts and bolts" required to make a new team function like a well-oiled machine in their current reality. They could not see any spare time in the day to sit in a room brainstorming productivity strategies when their reality was primarily consumed by survival tactics.

The potential existed to quickly head down a treacherous path of "point-less" meetings with "unnecessary" teams unless the senior and mid-level leaders came together on the role and outcomes that were only possible if a new team was formed. Not only would this take away precious time and resources, but it also ran the risk of eating away at the foundational aspects of the role of the team within the company and the important piece it played in the company's overall health and vitality.

And the team itself? Chances are good they would have been more like a group of colleagues sitting around a table than a highly functioning team.

How can you ensure that any team you form, or take leadership of, will truly begin its life operating as a team rather than a randomly assorted group of individuals? Establish a rock-solid foundation for teamwork from the beginning with these five foundational categories:

1. **Team life span.** What type of team is this? Is it a permanent, ongoing team that is integral to the operational needs of the company? Or a temporary team that has been formed for a specific purpose (such as strategic plan implementation) and has a specific dissolution date? Beware the tendency to form a team and then "figure out later" how long it should exist. Communicating the expected commitment from the beginning will allow team members (and their own leaders) to know if it fits into their schedule and if they can truly follow through.

2. **Team member roles.** Are there particular roles needed on the team? What goals are particular to each role? Are these roles and associated goals decided by the team leader or co-created with input from the team members? What method will be used to measure progress toward those goals and check in for shared accountability? A team naturally provides a space and place for diverse input, perspective, expertise, and experience. If you choose to embrace and champion this idea proactively at the start, your team will be set up for even greater impact.

3. **Meeting etiquette.** Will meetings start on time regardless of who is in the room, or will the leader wait until everyone is present? Will meetings end on time regardless of whether the whole agenda has been addressed? How is the agenda set? Can anyone contribute, or is someone the ultimate decision-maker? If an idea comes up mid-meeting and it is not on the agenda, can it be discussed? Meetings are an entire subject to themselves, but I bring it up under the concept of team leadership to drive home this point: Regardless of how formal or informal your meetings will be, they exist to advance team business and progress toward goals. An unwieldy conversation with no direction or a meeting that's called for an hour, then goes for two, is usually

a waste of time and could even have a negative impact on team culture. My personal favorite: meetings that are called to give the team an "update." The meeting and the message need to have a clear point, something the team needs to take action on. Do you want to pull people into a room, away from their job responsibilities, and pay them to sit and listen to general information? No. If this is really the case—if there really is "just an update" the team needs to hear—then that is an email. Not a meeting.

4. **Meeting speakers.** What is the process for inviting or allowing presenters at the meetings? If someone is delivering a presentation to the team, should that person use PowerPoint or handouts (i.e., a "deck")? Do materials need to be shared ahead of time so team members have a chance to review them? If your team can benefit from the perspective of an outside member at one of their meetings, make sure they understand the connection and why the speaker is there. Too often I've seen team members totally confused by new ideas offered by a colleague—should they take direct action on the idea? Or is it something to consider that can be rejected? And the speakers themselves need to understand the purpose for which they have been invited to the team meetings. Knowing specifically how their time speaking can be spent helping the team advance toward their goals is time well spent for everybody.

5. **Communication technology.** How will team business be conducted? Will a shared drive be used, or will notes be recorded in a certain file? Will members receive texts about various updates? Are computers and phones permitted during team interactions, or is the meeting intended to focus only on face-to-face time? The rise of technology carries with it the advantage of streamlining efforts and the disadvantage of split focuses. For every team member that is comfortable moving between technologies, there is a team member who is deeply set in their ways or even completely intimidated by using yet another form to collaborate. Do not let your team's productivity and interactions get derailed by

technology challenges. Take the extra time to make sure everyone is on the same page, comfortable, and engaged. Otherwise you could be one unread text message, inflammatory email, shared file malfunction, webcam mishap, underutilized mute button, or virtual meeting disaster away from the total meltdown of your team.

Building teams on a rock-solid foundation of shared purpose and operational norms will benefit not just you as the leader, but your team members and the entire organization. And whether you're dealing with organisms or organizations, this doesn't just unfold naturally. As the leader, *you* need to make sure the entire team is thriving on the same reef—from barracuda to rock shrimp to everyone in between.

Appreciate How Teamwork Holds the Organization Together

Rooting your team in common purpose and shared norms is the first step to ensuring that you build an effective team with a chance to achieve their stated outcomes. But to realize this desired impact, you must also develop and nurture this foundation as the living, breathing animal it is. It is crucial that, as a leader, you do not overlook the value of the individuals on your team, or the significance of the team within the organization.

I'm going to guess that at some time in your life, you've been around a team that really clicked, that hummed with purpose, that got things done. Maybe you loved how it felt to be a member of a highly functional team. Maybe you were the leader, and you were thrilled by how productive and symbiotic the team was. Maybe you were just an outsider watching from afar, but you still felt the impact of the team's work.

I'm also going to pose the opposite assumption: that sometime in your life you've been around a team that quite simply did not work, whether because people didn't get along, the parameters were ill defined, the leader was disengaged, or for some other reason. This team was just not the right fit, in any sense of the word. And again, maybe you had nothing to do with this hapless team, but you felt the impact just the same.

And I am going to hedge a bet that you've been a part of, led, or searched out a novel approach to one of the most ubiquitous, foundational elements of professional life: team-building.

Always intrigued by how I could combine leadership development with wildlife and wild places, I began drafting ideas for team-building events well before I was ever hired to do so. Once I began working in the novel setting of an aquarium, I was convinced it was a perfect out-of-the-box location to engage team members into flexing their creativity, brainstorming how to work together, and communicating more effectively. So when a multinational manufacturing corporation called to inquire about a day of team-building at the aquarium, it was as though all the stars had aligned.

"We want the day to be productive, inspiring, and fun," the woman on the phone said enthusiastically. "We believe in the value of team-building and know we need to continue nurturing this fundamental process, but we've also basically 'been there and done that.' So, in looking for a unique experience, we thought of the aquarium. What do you think?"

Ideas started pouring out of my mouth: what we could do for the team, how we could go beyond our training room, and why the aquarium exhibits and our animals would be so inspirational. A few weeks later, two other staff members and I were leading the corporate team toward the coral reef exhibit for the first team-building exercise of the day.

"Which animals catch your eye first?" I asked. "Which ones do you want to follow a little more closely and learn more about? What is the first animal you want to tell your family about when you go home tonight?"

People called out their answers, fascinated by the sharks with their commanding presence . . . the sea turtle that seemed to be heading in no particular direction . . . some sort of fuzzy creature with long antennae . . . two eyes barely visible within the rocks . . .

Then I asked: "What about the coral?"

Silence.

"What does the coral provide for these animals?"

The answers started coming: a place to hide, shelter from sunlight, a place to attract sea life that other animals might want to eat . . .

"And what if the coral went away?"

More answers: The animals might lose their source of food, their community, their shelter . . .

"Did you know that coral is *alive*?" I asked. "That each one of those large structures you see in the tank is actually made up of hundreds or thousands of individual animals?"

Everyone's eyes remained glued on the tank in front of them. No one had noticed the largest community of organisms represented right in front of their eyes, the proverbial elephant in the room—well, the coral in the tank. Looking at all the "pretty" fish, they'd forgotten the coral was alive too. They'd taken for granted the *one animal* that was anchoring the entire ecosystem.

"Are you sensing the point I'm trying to make? The largest, most important component of a well-functioning team is very likely the one most at risk of being overlooked or taken for granted. Maybe it's the tendency to overlook foundational aspects of communicating a shared purpose or operational guidelines for the team, or maybe the individual people on the team are more introverted or therefore not as recognized or noticed. But just like the coral, if your foundation is overlooked, everything else falls away. And sometimes this happens so quickly, you won't even notice it until it's gone."

Murmurs made their way through the team of people as they gazed with new understanding at the team of coral.

What about you? Would you have noticed the coral right away, or would it have blended into the background as the other organisms captured your attention? Would you have guessed that the reef in front of you was made up of thousands of tiny animals—the coral itself?

Well-functioning teams are the coral reef of an organization: the living, breathing structure that ties it all together and allows all the moving parts to work and thrive. And like coral, some teams, or even individual team members, go unnoticed or underappreciated at times. Nonetheless, they remain the necessary foundation on which a highly complex ecosystem can function and flourish.

You can draw inspiration from the fact that you're starting from a place of appreciation: The need for a team has arisen, and you have decided (or signed on) to help form one. Get excited about the prospect of moving the needle in ways that only a group of collaboration-minded people working together can. Remember that your team members can provide unique perspectives that will positively impact your goals. Taking the time to seek their input from the very beginning will strengthen their sense of purpose, encourage their buy-in to team operations, and boost the impact of their teamwork.

Although team members may see things from an entirely different vantage point, perhaps they are just as excited as you are to discover how working together can contribute to the bottom line and the company's future.

Beware of Erosion to the Team Environment

Learning about and participating on teams is one of the earliest lessons in leadership we receive. Think back to your childhood and grade school, when you were involved in family meetings, reading circles, team projects, and group assignments. I remember singing with chorus classmates, passing the ball on the soccer field, working together on toothpick bridges, and getting out of the way so my doubles partner could return the tennis ball. And I distinctly remember times when team members functioned together merely as a group—not as a team.

Those uncomfortable memories are seared into my brain. Because when a "team" operates as little more than a group—without universally accepted guidelines, a sense of collaboration, or mutual appreciation for all members—the impact is far-reaching and longer-lasting. If things break down, it takes extra time, energy, and trust to rebuild that group of people back into a true team—at least double the time it took for teamwork to erode in the first place.

Coral live in one spot their entire life. They grow the reef by splitting and regenerating, as well as by spawning (releasing) eggs and sperm into the water that will join and develop into larvae, then drift with the ocean currents and settle in new locations. It takes a long, long time to build a reef.

Coral are good neighbors, too. They're intimately linked to the nearest members of their coral community by a layer of mucus that allows them to communicate and protect against infection. If disturbed, however, bacteria or other disease can penetrate this mucus layer, which can lead to disaster. When normal operations are interrupted for too many of these organisms, the entire reef can become sick and die.

Now I want to take you back underwater with me, on our dive in the warm water to where the coral lives. We noticed the big animals first, the black tip shark and the remora swimming by. Then we caught a squir-relfish and a queen angelfish checking us out. But now we finally notice the reef itself, and the sheer enormity of it is unanticipated. So we swim closer to take a look. We might even be tempted to grab on to something to help us stay in place. Something that can help us observe just what that tube worm we can see out of the corner of our eye is doing. Then we might want to stand up for a different perspective, so the tendency is to reach out to grab on with our hand or put down our foot straight onto the coral.

This is sheer reaction to such a solid structure: We lean in to it and use it to help us. But as we know now, this sturdy, awe-inducing structure is more than solid rock. It *is* that—literally made of limestone—but it is more. It is *alive*. And as commanding and strong as it might appear, each tiny bump of a flipper, change in water temperature, or chemical pollution from sunscreen compounds on itself and begins to erode this solid foundation. These minute changes may not be seen at first—in fact, many times it is only after the erosion and damage has reached a systemic level that the true impact can be seen throughout the reef.

Once this destruction begins, it won't be long before a similar disaster strikes other species. Coral reefs are called the "rainforests of the sea" because of the diversity of life they support. Approximately 25% of all marine life, at some point in their life cycle, depend on coral reefs for shelter, food, and a place to reproduce and keep their young protected.[4] And up to a half a billion people depend on coral reefs, too, for food, protection from oceanic storms, tourism, and fisheries.[5] Without this foundation, these organisms have no structure to depend on and cannot thrive.

Never underestimate how quickly things can fall apart for a team. When seemingly innocuous behavior, lack of adherence to agreed-upon guidelines, or even a clearly uninvested team leader goes unchecked for a while, the surface reflection may not accurately represent that full impact: The long-term damage has already been done.

A healthy coral reef provides the integral foundation for other animals to thrive, with food, shelter, and space. It grounds a beautiful, strong, flourishing ecosystem—one that operates purposefully according to nature's laws, one that can adapt to external pressures and endure. Likewise, a healthy organizational team grounds a purposeful, strong, robust organization—one that operates with purpose according to established guidelines, can flex with external changes, and will stand the test of time.

Take away the community of coral—whether intentionally or accidentally, by not appreciating it or even realizing that it's there—and there is no more reef. All the animals will leave. An entire ecosystem will be destroyed.

Take away any foundational aspect of a team—whether by assuming everyone knows the team's purpose, operations, and guidelines, by running out of time or energy to put this foundation into place, or by not fully valuing each member's individual contribution—and there is no functioning team. All the people will flounder in extraneous effort. An entire chunk of time will be wasted.

Your company does not need another group.

They need a team—with a foundation built on solid rock.

Period.

Unbreakable Law #6 Pro Tips

➤ Teamwork starts with a rock-solid foundation of what all teams need in order to survive: a shared understanding of their mission, purpose, goals, and operational norms.

➤ Great leaders recognize their teams and team members as the living, breathing support network that allows the organization to thrive, much like a coral reef supporting an entire ecosystem.

➤ Building and nurturing an effective team is a long, intentional process, yet disruptions in the environment can erode that foundation in the blink of an eye.

There is no one-size-fits-all approach to teamwork.
Surround yourself with people who are not like you.
And never underestimate the impact of a solid foundation.

You are a leader of people.
Each one is unique and able to fill a certain niche on the team.
When a team is built on solid rock, its impact can reach all the corners of the globe.

Behind the Scenes with KiwiE

Our aquarium leadership was very interested in having mullet, a common schooling fish, in the shark exhibit. The staff were skeptical, as these species of sharks very much liked eating mullet.

The first time a bunch of mullet were added to the shark exhibit—they were eaten within a few moments.

The next time, the sharks were fed directly before adding live mullet to the exhibit. This time, the mullet lasted a few minutes longer—but the outcome was still the same.

On the third attempt, the staff fed the sharks even more food before adding the mullet. But still, no mullet outlasted the sharks.

The leadership team eventually acknowledged that mullet could not live in the habitat with the sharks. Yet we had made an unintended discovery: The visitors really enjoyed watching the sharks feed! It was a behavior most people never get to see in the wild.

Today, mullet still do not live in the shark habitat. But the sharks have now been trained to come to various stations around the exhibit where they are fed, all under the watchful eyes of curious aquarium visitors.

INSTINCTUAL LEADERSHIP FIELD GUIDE: TEAMWORK

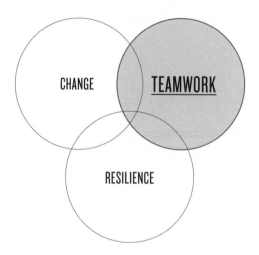

Now it's your turn! I invite you to reflect on the ideas, tools, case studies, and insights provided in the previous three chapters, all centered on the leadership concept of *teamwork*. The following prompts will guide your reflections as you critically evaluate the relevant lessons and how they can best help you.

To target your answers and make them as immediately applicable as possible, I advise you to think about your answers only as related to your current role as a leader.

As you're reflecting, think of your own examples, case studies, and maybe even a few animal stories! Jot those down in the white spaces around these prompts. They will come in handy as you refer back to this book and pass along these lessons to others in your life.

This is how you truly personalize and implement the wisdom from the wild as it relates to teamwork.

1. The idea that resonated with me the most from Part 2 is:

2. The biggest opportunity regarding teamwork I can see is:

3. The biggest challenge regarding teamwork I foresee is:

4. I am most excited about:

5. I will take action and hold myself accountable by filling in one idea in each box below, building off of the ideas in Part 2, to be implemented in the next three months:

STOP	KEEP DOING	START

6. I will pay it forward and help others whom I lead and influence by:

_____

_____

_____

_____

PART 3

RESILIENCE

"LEADERS ARE THE ONES WHO HAVE THE COURAGE TO GO FIRST, TO PUT
THEMSELVES AT PERSONAL RISK TO OPEN A PATH FOR OTHERS TO FOLLOW."

—Simon Sinek, author, speaker, optimist

Resilience.

The reality behind what it takes to lead.

The fundamental aspect of showing up at your best so you *can* lead.

Arguably the most overlooked and underappreciated component to leading at your highest level—to leading from the inside out. Many leaders keep going, and going, and going, far beyond what they expect of their team (and sometimes far beyond what they know is feasible). Leaders are the ones who keep showing up, the people who stand alongside, in front of, or behind their team not just when it is easy, but more importantly, when it's hard. True leaders are the ones who make decisions, learn from each one, and come back the next day for more.

But some leaders ignore the signs of exhaustion and continue to push beyond all reason, betting on the fact that rest will be just around the corner and their resilience bucket will be refilled. They believe they can continue to outrun themselves, time and time again, and can continue to adjust and deal with the consequences as they come, well into the future. They are wrong.

When working with my clients, leaders have told me that they are "juggling too many obligations" and "struggling to find the time to prioritize big picture and strategy, while being bogged down with the small, but necessary stuff."

These clients have even started to feel the pressure related to their leadership skills. They explain to me that they "worry too much about being 'liked,' which at times sacrifices my instincts for change or new direction." One leader shared with me that they "constantly feel I'm just holding on. A small part of me thinks this must mean I'm not supposed to be here doing this job; I must be an imposter. How do you silence the voice of doubt in the back of your head?"

Above all else, leaders know that they are consistently aiming to balance priorities at work, lead their teams effectively, manage their energy, and listen to their own needs for resilience. As one hardworking leader told me, they are doing all of this while also "striving to walk the talk and set the example for others around them."

This is a tall order, and leaders who do not refill their own well—failing to replenish themselves in mind, body, and spirit in a way that will support their purposeful leadership journey—will fall short. Eventually, their strength will give out in all ways possible, and they will not fulfill the leadership purpose for which they were destined.

Your leadership journey is not bound by any particular career choice or desired title, nor is it contained within any set decade of your life or project you're working on. Sometimes the role of leader is designated by someone else, and sometimes it is assumed organically, even by someone not exhibiting stereotypical leadership qualities. Leaders emerge on the playground in elementary school when a group of friends is trying to decide which game to play. Different leaders take charge when an unexpected catastrophe strikes and the previous leader is suddenly unavailable. And a newly successful leader may materialize when every other strategy has been tried and a fresh idea is the only way forward.

Leadership is a combination of your *miles* of experience and the *moments* that you are confronted with. These two ingredients provide you with the individualized frame from which you lead as only you can. Combine them with an awareness of the centralized role resilience plays in effective leadership, and you become unstoppable.

Why is it so important to embrace resilience and put it into practice as a leader? Because driving and surviving change is not a one-off game. Because building and leading an effective team is always a different journey than the last time. And the people you are leading are watching you. They take their cues from what you do, not what you say. So, if you expect them to flex their resiliency muscle, you must flex yours first.

Resilience is a biological certainty based on the fact that humans, like pelicans, sea cucumbers, and cheetahs, are animals too. I take heart in this truth, as I believe it negates all potential controversy over the need for downtime, for self-care, for recovery. It is a biological fact: We need rest in order to be fully human, let alone be able to lead at our highest level. Leaning in to and championing the concept of resilience ensures we can continue to show up—and inspire those around us to do the same—as we all step fully into the leaders we were born to be.

Why now? Why again? And why will I have to continue to embrace this?

Because leaders are built over miles and moments.

Behind the Scenes with KiwiE

When I rented a car in New Zealand, I was casually filling out the expected paperwork when I noticed an unusual clause: I needed to sign and agree that I would not purposely put the car in a situation in which keas could potentially eat the rubber on my car.[1] I'll admit that I've signed so much paperwork over the years that I typically just go along and sign what I need to, but this time—can someone please explain what a kea is and why it might eat the rubber on my car?

I would learn that keas are large alpine parrots that are quite cheeky in nature. They had indeed been known, at times, to eat the rubber that sealed the window of a car door. I would also learn later, while hiking the famed 33.2-mile Milford Track up and over a mountain and crossing 200+ potential avalanche spots in an area that averages 182 days of rain annually, up to 30 feet, that when stopping to sleep overnight in the huts it was advisable to leave your hiking boots outside to dry, but to keep them well out of reach of the keas. They have been known to take a hiker's boot with them up to the top of a tree and produce a call that sounds suspiciously like they are laughing and mocking the poor person stranded below without a crucial hiking boot, in the middle of a multi-day hike.

So, what would my fate be with keas? I'm delighted to report that I did get to see them in the wild and respect them from afar. I also happily made it out of New Zealand with the rubber around my car doors intact—and in possession of both of my hiking boots.

RESILIENCE IS INSTINCT IN ACTION

(Just Watch the . . . Pelicans)

Which bird comes to mind when you hear the word "leadership"?

Is it the emperor penguin, whose loyalty is displayed by the male standing guard over the family egg for weeks on the frozen tundra, slowly losing body fat while patiently waiting for the female to return from a hunting trip?

Or the peregrine falcon—the fastest bird on the planet—whose singular focus is demonstrated in its steep, close to 200-mile-per-hour dive toward its unsuspecting prey?

Or the belted kingfisher, whose short, stocky bill is perfectly adapted for catching small fish along the shorelines of lakes, rivers, and streams?

I'm going to take a "wild" guess and assume that the bird you thought of first is the one most often associated with regal traits of power and leadership: the bald eagle. And of course, what American does not feel pride in the bald eagle, our national bird, which presents such a striking figure with its golden beak and white-feathered head, soaring majestically over land and water?

But look closely enough, and you'll start to see that certain traits of the bald eagle are not as admirable in, or expected from, a leader—like the tendency to steal food rather than catching its own, or its susceptibility to being attacked by smaller birds. As the story goes, even Benjamin Franklin felt the turkey would be a better choice for our national bird than the bald eagle, which he called "a bird of bad moral character" and "a rank coward."[1]

No one, not even the bald eagle, is perfect.

But it's not the idea of perfection we should be after. No animal fully embodies all aspects of leadership, any more than a single person does, but we can still take away nuggets of learning from each example. And as you know by now, I'm far more interested in finding inspiration in the unusual places—in this case, in a bird that may not first come to mind when you think about leadership.

Like the brown pelican.

All over the coast of southern Florida, in the same habitat as seagulls, ospreys, and skimmers, you can find brown pelicans. They are easy to spot: Just look for the birds that look like they shouldn't be able to fly, with their seemingly heavy bodies and large beaks. But the funny thing is, they *do* fly. Over and over again, they fly and dive, fly and dive, fly and dive after fish in the water, searching for something to eat, and while they are not always successful, eventually they will be. And when they're not flying and diving repeatedly? Chances are you'll see them sitting by themselves on the water.

During a college-level course in ornithology, when I needed to study a species in the wild for my final project, I immediately chose the brown pelican. It could have had something to do with the unwelcome prospect of standing outside in snowy Ohio netting songbirds, when instead I could be on the beach in Florida, watching pelicans.

So, I announced to my family and friends in Florida that they had been recruited to study brown pelicans with me during spring break. Then I ignored their side-glances and gathered them on the beach. Turning my head away from the curious looks of beachgoers eyeing our large spotting scopes, clipboards, binoculars, and stopwatch, I pretended not to care that people were staring at us, trying to figure out what we were doing. Instead, I focused on the task at hand: learning more about how brown pelicans spend their time.

And how they embody resilience as instinct in action.

Embracing Resilience Should Be Standard Operating Procedure

Throughout my career, I've come to understand that I can intentionally effect change and lead by example for those who are watching and learning from me—regardless of my current industry, job title, life phase, or leadership role—by embracing resilience. In fact, I recommend making resilience a standard operating procedure for leadership. And here's what else I have come to know through my journey as a leader, while both leading others and watching those around me lead: that at its core, resilience is simply instinct in action.

Has there been a time in your life when you have felt on top of everything? How often do you really feel that it's all dialed in—a handle on your workload, a consistent morning routine, knowing how to manage a nutritious diet, exercising on a consistent basis, laser-focused on your next career milestone? I've had times in my life like this, when everything was flowing, my creativity was off the charts, and I took quality time for myself. I really knew then how to go to work each day and perform at my highest level.

Has there been a time in your life when everything felt like it was coming off the rails? How many times have you felt totally unproductive—unable to figure out why your team was imploding, hit by sudden change from every direction, needing more sleep than you're getting, clueless about the last time you ate any fruit or vegetables, and with no idea where you left your running shoes?

I've had times in my life like this, too, when I don't even bother to open my computer or phone to catch up on work and get on top of my email inbox . . . when I'm dazed by the parade of people entering my office with a complaint to share, a new idea to try, a project they're behind on, or a strategy that the higher-ups want us to consider . . . when I wonder whether anyone will notice if I even came into work that day . . . when I wonder whether anyone will notice if I just sleep in my office, because I have no time or energy to drive home and come back again the next day. Perform at my highest level? Ha—I can't even remember what that felt like or when I last experienced it.

Every one of us has had periods of time like these. Thankfully (hopefully!), no one is destined to stay on the negative, overwhelming side forever. But days can quickly turn into weeks, which can turn into months, and suddenly we can find ourselves wondering where all the time has gone while we've been struggling to achieve, make a difference, or just stay afloat.

It's easy to get caught up in the momentum, whether it's a winning streak or a downward spiral—and that's when you should dial in to what you alone need to nurture *your* resilience. Not what your team needs. Not what your CEO, your friends, or your mentors need—what *you* need.

There is risk in overthinking the concept of resilience, however, and simply adding it as yet another item on your to-do list. An eager leader might subscribe to too many newsletters that explain, in only five simple steps, how to relax and take care of yourself. Likewise, signing up for too many wellness retreats could lead you to believe you don't know how to make decisions that are best for you. Moderation in all things, including stress management and self-care routines, is a virtue. Pack your schedule with exercise, meditation, juice cleanses, detoxes, gratitude journaling, scheduled time with friends and family, time to yourself, time outdoors, time with your pets, and more . . . and suddenly your "downtime" for recharging is overscheduled—just as jam-packed as your work calendar.

There is also risk in aligning resilience too closely with relentless positivity. Becoming resilient does not mean ignoring struggles, despair, or difficult moments. It does not mean forcing yourself into a happy state and charging into the next project. Task yourself with the impossible goal of being positive and upbeat all the time, and you will quickly learn the toll it takes on resilience and mental health, both yours and that of the people around you.

Resilience is about being able to keep moving forward, in the way that only *you* can. Your capacity and needs surrounding the concept of resilience are unique to you. By leveraging the insightful research and tools available, you can assess your own needs and build your resilience muscle.

Leadership development has come a long way in incorporating resilience as a skill that is integral to leadership success. Not all that long ago, mental health issues like resilience weren't openly discussed in public, and bringing it up in the workplace could have been considered unprofessional. But now, organizations are learning to proactively embrace and champion the reality that for leaders and their team members to thrive, resilience is a necessity.

Throughout my career I've been fortunate to lead a wide range of teams. I've also had the huge honor of being responsible for other people's children. And I've had the amazing opportunity to learn how to respect, handle, and take care of wild animals. Put together, these components— colleagues plus children plus animals—form the perfect trifecta describing my career. If any part of this leadership triangle seems threatened, overworked, too excited, or simply in need of attention, I do everything in my power as a leader to help.

One of my teams was full of gregarious, outgoing individuals, responsible for both kids and animals, whose enthusiasm brought just the right attitude to the job every day. They worked hard and played hard. They had crazy, inconsistent hours and interacted with a wide range of people and animals who depended on them. Truly, there was no "typical day" for this team, which made it an exceptionally challenging and yet personally motivating team to hire for. Whenever I did strike gold, finding a team member who complemented the others and enhanced the work of the group, it was a magical moment.

One afternoon on my scheduled day off, knowing that we were in the middle of an important project, I arrived unannounced at the office.

Instead of the usual carefree but hardworking attitude I normally encountered, the air was tense and silent. The few team members present were either noticeably buried in their work or avoiding eye contact with me. This drastic change stopped me in my tracks. I stood there waiting for a signal, anything that would cue me on what was wrong and how I should react.

One of my team members rushed toward me, his usual grin replaced with a grim face and worried eyes. "I don't even know where to start," he began urgently. "Some of the team members, when they came in this morning, were clearly off their game . . ." He went on to explain that three key players were overtired and overstressed after a string of long, exhausting days working on the big project. The previous night, in an attempt to let off steam, they had stayed out too late, which only added to their lack of attention and focus today, especially since they were expected to be responsible for both kids and animals at the same time.

The team member in my office was beside himself with worry—for everyone. For the three key staff members, who were clearly overworked and not taking care of themselves. For the remaining staff members, who now had to pick up what the others couldn't handle. For the animals and the kids, who required our complete focus and undivided attention. For himself, trying to figure out his place as a senior member of the team in this situation. For me, having to worry about and wonder what potential repercussions this situation would cause if we didn't get a handle on it immediately.

Because the problem in this situation wasn't just that a few staff members were tired. It was also the fact that their actions were starting to sow deep rifts and lead to life-changing consequences such as becoming suspended from or losing their jobs altogether. I could tell they couldn't fully understand how far-reaching the outcomes could be, how comprehensively it would affect their career, and how devastating it would be for them to try to bounce back.

For me to address this as a leader would take a whole-person approach—it was not enough for me to simply address the obvious behavioral issues and risks to the rest of the project and to the team. I had to dig deeper to address the root cause of this behavior—was it within our department? The team? The project? The expectations? The training? I would be holding them responsible. But I also wanted to help them embrace their instinctual need for resilience as they were going to need this trait in the days, weeks, and months ahead.

You are a leader who is different from your colleagues, your mentors, your friends, and anyone else. We all have different needs, different ways of dealing with stress, and different levels of tolerance. We have at least one thing in common, however: To lead at our highest level, we must view resilience as a standard operating procedure in our leadership toolkit. Good leaders know when to forge ahead, when to take a break, and what they uniquely need in order to regroup before pushing onward.

Which brings us back to the pelican.

There I stood with my family and friends on the beach, clipboards in hand, charting the behavior of pelicans in set-time increments. When the stopwatch went off, we would all look to our designated bird and write down what it was doing—usually flying, diving, or sitting. Occasionally it was swallowing—but only water, or a fish? We quickly learned the difference.

If, after a spectacular dive toward the water, the pelican sat with its beak pressed up against its body before slowly raising its head, it had not caught a fish. It was simply waiting for the water to drain from its beak before it could fly again. We labeled this as "not successful" on our spreadsheets.

But if the pelican sat for a bit with its beak pressed up against its body before suddenly throwing its head up into the air backwards, it had caught a fish! It first had to drain the water from its beak before it could flick back its head and swallow its prey. We labeled this as "successful."

And regardless of whether a bird was "successful" or "not successful" in terms of catching a fish, there it sat on the water—sometimes for a few seconds, sometimes a few minutes, but the effect was the same. It was taking the time it needed to regroup and gather energy for the next go-around.

Sometimes the dive to catch a fish didn't go so well. It seemed as if the bird hit the water at an angle it wasn't expecting or there was some other interaction that caused the pelican to sit a bit longer than it had before.

WISDOM FROM THE WILD

Sometimes other birds, swimmers, or even boats started encroaching on the pelicans and I would notice changes in their behavior.

I like to think these pelicans were tapping into their own resilience by relying on instinct for what they needed. Not all of the birds we were watching exhibited the same behaviors. Some sat longer on the water before flying again. Some reacted quickly to an approaching boat by flying away, and others flew right toward the boat. Some dove again and again and again with only a few seconds in between, and one bird just sat on the water during our entire observation period.

Why? As a budding scientist observing their behavior, I was only capturing a fraction of their day. I had no idea what their behaviors had been prior to our arrival, or what they would do after we packed up for the day. What I was observing and recording in the moment was everything I could see right in front of me, and then I interpreted it through my research-informed, yet ultimately human, lens. What I could see happening in the moment was that each bird was behaving as they were instinctually designed to behave. And that each bird was relying on their instincts to tell them what they needed to do to move forward when things didn't seem to go as planned—whether that was flying away from a boat that was coming their way or not bothering to dive for food and instead just sitting on the water.

Resilience is a moving target that shifts with every season of life. And it is arguably the one component of leadership that can torpedo all the others, if not given its due. You may be able to limp your way through a team meeting or wing it during a client presentation, but if you can't get out of bed in the morning with a clear head, the determination to move forward, or enough energy for the day, nothing else gets done.

Lead Yourself from the Inside Out

You know yourself best. You know instinctively what you need—at this phase of your life—to be resilient. And to be a resilient leader who can lead others? You must lead yourself first. You must listen to your own instinct.

This doesn't mean we can always hear this voice inside us, pay attention to it, or even feel as if we have the time or permission to do so. Nonetheless, you *do* have an instinctive voice inside that knows what you alone need.

Understanding and embracing instinctual leadership as a concept must incorporate the study of resilience. Of course, we can learn from great leaders who have gone before us. We can read the latest literature and research papers on leadership and develop relationships with mentors we trust. But all too often, we overthink leadership and rely on the wisdom of relative strangers rather than trust that we know in our gut what to do when it comes to resilience.

Resilience is not a list of self-care activities or ideas gathered from other people. It is a muscle of your own that you can develop, flex, and keep in shape. And that muscle's capabilities change throughout your life, with each phase, each job, and each reality you find yourself in.

I am a completely different leader now from when I was single, in my twenties, living in an apartment in a big city, responsible only for me and my two cats. I worked on my days off, went in at all hours of the day and night, and would drop everything and travel to any destination where I was needed. I was always up for after-hours events, speaking at conferences, and joining local professional groups. I rollerbladed to work, took exercise classes at night, learned how to cook for myself, and didn't need much sleep. As I faced any obstacles or challenges head-on, surrounded by like-minded colleagues in similar life phases, I bounced back full of confidence and maybe a touch of bravado. I was a leader with a "roll with the punches" approach to resilience, which was in line with my instincts at that time and place.

I am also a completely different leader now from when I was pregnant with my first child and responsible for a team of forty-five people. At that time, I could feel how the obstacles and challenges were now taking more energy and attention than they were before. I had more viewpoints to consider, farther-reaching consequences to be mindful of, more people I was responsible for, and a broader responsibility both to my company and to my growing family. My time was now occupied by conference calls on the hours I was commuting in my car, and books I now listened to on tape. I hardly ever saw my friends outside of work, and I couldn't remember when I'd last had a massage. I was a leader employing the "which fire is most important now?" approach to resilience.

I am a completely different leader today from the leader I will be for my next client, my next team, my next committee, my next board position. Some aspects of life I have learned to consolidate and can respond to quickly before moving on. I have learned the wisdom that only time can provide: which opinions I should take to heart, which feedback I should incorporate, and which comments I should let go of completely. I have learned which experiences deeply affect my energy and how to maintain my stamina over the years, which is a perspective that can only be gained through years of experience.

One of the most surprising aspects of resilience I've discovered is that its importance is independent of industry, job category, organizational structure, or even tax status. An individual person's way of exhibiting and nurturing resilience may differ. But the imperative for being resilient is consistent.

As a leader in nonprofit organizations, I led teams that worked hard, sometimes for more hours than was really healthy, and believed deeply in what we did. We commiserated when we were tired, and I spent many hours building up my team, reassuring them that they mattered both to the mission and to the bottom line, and that they were understood and appreciated by other departments. Their resilience was deeply connected to believing that their integral role in the organization was recognized and valued, so when we felt overworked, I would continue to motivate them and support their individual needs.

As a leader in corporate organizations, I led teams within the same job function as I had in the nonprofit world. I had assumed that my leadership strategy would be different now that I was leading a team with a higher pay scale than my nonprofit teams. I had assumed some of the inherent struggles of feeling like they were valued would be gone since they were compensated differently. Much to my surprise, I soon discovered that my team faced many of the same challenges and voiced similar struggles.

They worked the same long hours, and I still had to help them manage their time. They also wanted to matter, both to the mission and to the bottom line. They did not want to feel taken advantage of or misunderstood. They, like me, believed they were integral to the organization's success and wanted to be valued for their hard work, not just for being a revenue-generating department. They needed this validation so they could bounce back when challenges came their way.

The experience of managing and leading these teams through such similar struggles and opportunities reinforced one basic concept in my mind: No matter the type of organization you work for, we all need to build and maintain our resilience muscle. Our problems are not solved solely based on the amount of money that we make. And we maintain that muscle by tapping into our instinct and leading from the inside out.

We are born with certain traits that will help make us the leaders we're meant to be—including resilience—but we need to learn how to develop them fully. We grow and learn to become more effective leaders over time by tapping into what makes us unique, and then taking action.

My two children are the lights of my life, and I learn from them every day. Tasman literally came out of the womb speaking. She has been a communicator since before she had words to use for her thoughts. She is the first person to volunteer, loves performing onstage and will always try out for the solo, and immediately takes charge in every group project. She is strong and confident, and will speak up in any situation, especially in giving voice to those who feel unable to speak up for themselves. While I was laboring with her little brother, Tasman was in the hallway instructing and leading my husband, my father, and my mother in yoga—at not quite eighteen months old. I am immensely proud to be her mother.

People have always told me that my daughter is a natural-born leader. Maybe it's because she displays leadership skills that are outwardly appreciated and usually aligned with what many of us assume a leader needs to do. I agree with them 100%—she *is* a leader, and I'm excited to see how she will use her instinctual skills and develop them fully to become the leader she is meant to be. I'm also intrigued as to what she will identify and employ for her resiliency strategies as she continues to grow as a leader.

Is there such a thing as a "born leader"? Are we created with an innate sense of being able to lead, or is this a learned skill? And if you are a born leader, what does this really mean? Perhaps you instinctually know how to motivate people to action or give a persuasive speech. Maybe you are a whiz at managing a group of people or putting together and executing a viable plan for change. Maybe you know exactly how to get to the heart of what really matters so time is not wasted. And all while maintaining a balanced budget and a balanced life.

It doesn't matter one way or the other. Each of us is born with unmistakable qualities that make us who we are—that make our instinct uniquely ours. When we listen to the instinct, lean in to it, and take action, we grow more fully into the instinctual leader that we are and unleash the resilience we need to lead at our highest level.

Find and Build Your Resilience Muscle

How can you tap into your natural-born traits and become an even more resilient leader? Good question. Let me teach you two tools that will show you how to discover the answers yourself.

Characteristics of a Resilient Leader

Get out a piece of paper; a small piece or a page in your journal is fine if you are working as an individual. Or if you are leading this exercise for a team, use a large piece of notepad paper that can be stuck on the wall, so you can write down answers and facilitate conversation.

Here is the question:

What are the characteristics of a resilient leader?

Be sure to answer off the top of your head, with no editing and no feedback—just in-the-moment brainstorming. Consider a time when you saw this in action from a leader you know, or remember a time when you exhibited this trait yourself. Keep making a list until you have at least ten characteristics, or until the ideas from your team start to diminish. Make sure you wait until the end, and then take a step back and reflect on the answers.

Here are some typical answers:

- Saying no

- Giving others space for what they need

- Not covering up difficulties with positivity

- Treating yourself as you expect others to treat you

- Leaving work on time

- Only taking on as much as you can handle

- Communicating when you need a break

- Respecting the mental health needs of yourself and others

Now, as a follow-up to this exercise, ask yourself (or the group) a second question:

Which of these characteristics do you have direct control over as a leader?

Circle all the ones that you, as the leader, have direct control over. Be fair and honest as you consider each response and decide whether or not to circle it. In my experience, if you probe deeply enough, at least 90% of the answers are in fact characteristics that you have direct control over in your capacity as the leader.

What's the takeaway?

If you want to be a resilient leader, you need not look to others to make this happen for you. Nor can you! Because nearly all of the characteristics you have identified that make a resilient leader are entirely within your control.

We often believe (or are told) that we cannot be resilient because of what people impose upon us. We feel the weight of other people's expectations, deadlines, or influence as to how we should act and respond. But the truth is, there is room to find and build your own resilience muscle. In fact, you have just worked through a brief exercise in which you identified what makes a resilient leader and what you have control over.

Now it's time to put the results into action.

And how do you do that? By becoming aware of, more involved with, and more responsible for how you spend your time and, thus, your focus and energy.

Using an Ethogram to Explore Your Resilience

Remember all my time spent on a beach observing pelicans? Did reading that story make you wish that you, too, were on a beach—maybe not actively studying pelicans, but soaking up the sights and sounds of the ocean?

This next tool borrows directly from my pelican research and will help you get a realistic picture of your current spread of focus and energy to better understand how you can incorporate strategies for nurturing resilience. Some leadership development courses call it a time study, but in scientific circles (and in scientific research), this tool is called an ethogram.

The exercise to complete an ethogram will help you take a step back and objectively analyze your behavior, just as a scientist would. It will help you adopt a nonjudgmental mindset as you see the big picture of how you spend your time, focus, and energy—no questions asked—and proactively decide how and when you can incorporate more, less, or different behaviors to help you employ resilience.

The ethogram is designed to help researchers categorize and understand patterns in behavior, without it becoming a never-ending, pointless list of "things the subject was doing at the time." Here's how you can use it for resilience:

First, make a list of the following:

- The first ten things you did when you started work yesterday

- The first ten things you did after lunch

- The first ten things you did before concluding work

- The first ten things you did before heading to bed

- The first ten things you did over the weekend

Next, look at your list of these fifty things and create broad categories that relate directly back to the characteristics you described in the first exercise about what makes a leader resilient. Don't be too hard on yourself—just like the SWOT analysis, this is not a perfect science. The categories will be different for everyone, but here are a few examples:

- Technology

- Face-to-face interactions

- Physical health

- Mental health

- Socializing

- Strategizing/visioning

- Problem-solving

- Busywork

- Fun!

- Me time

- Family time

Now that you have your categories, you will be charting your behavior over the next three days. Decide which three days of the week make the most sense given your professional role. I recommend at least two days of full-time work and one day when you are not working.

Create a spreadsheet that looks like the example that follows, listing your day in thirty-minute increments. You can drop down to fifteen-minute increments, but that approach can be overwhelming as it can result in too much data. Forty-five-minute or sixty-minute increments, on the other hand, give too little detail to be effective.

As you go about your day as usual, fill in your behavior during that time frame and add any brief notes that may help you understand or interpret it later. Do not fill in the Category column until after you have completed your observations of your behavior for the full three days. When you are actively completing your ethogram, thirty minutes at a time, you are simply entering data into the Behavior and Notes columns.

TIME	BEHAVIOR	CATEGORY	NOTES
7:00 – 7:30 a.m.	Eating breakfast		While reading the news
7:30 – 8:00 a.m.	Exercise and meditation		
8:00 – 8:30 a.m.	Checking email and voicemail		
8:30 – 9:00 a.m.	Checking email and voicemail		
9:00 – 9:30 a.m.	Leading a team meeting		Remote team meeting

Resilience Ethogram Example (Day One)

After you have completed three days of observations, go back and fill in the Category section based on the categories you brainstormed earlier. If a certain behavior doesn't quite fit into any category, feel free to create a new category—but try not to create so many categories that you end up with a mishmash of observations. The goal is to identify patterns and to evaluate how you are spending your time, focus, and energy. Because, as the old adage goes, "where your attention goes, your energy flows."

Next, on your completed Resilience Ethogram spreadsheet, highlight the rows that you believe relate directly to resilience.

Now ask yourself—without judgment or expectation—what do you notice? Are you flexing your resilience muscle as often on a workday as you are on your off day? Are you using resilience at various points during the day, or only at certain times (e.g., morning, evening)? Is there a need and/or an opportunity to rearrange your schedule in order to proactively add more resilience-building activities to your day? Is there enough time, focus, and attention on resilience overall?

Here is the final step: You can now use all the tools in front of you to fully reflect on your instinct and purposefully strengthen your resilience muscle. Comparing the first list of resilience traits in a leader with the pattern you have observed on your ethogram, I invite you to identify *one* strategy you can reliably integrate into your schedule that will move you along the path toward becoming a more resilient leader.

Some strategy ideas include the following:

- Insert a weekly fifteen-minute check-in with a mentor for accountability and feedback.

- Designate screen/technology breaks each hour, to give both your mind and your eye muscles a rest.

- Identify a thirty-minute period that you can spend outdoors each day (bonus points for a consistent time slot each day, so your mind, body, and spirit can learn to expect this rejuvenation).

- Isolate an activity on your ethogram that you know takes extra energy and focus and build in the time you need to "ramp up" and "ramp down." (For example, don't schedule back-to-back meetings that could unnecessarily zap your energy and focus.)

- Schedule a regular thirty-minute "resilience activity" on your shared work calendar, to help you serve as a resilience role model for those around you.

Remember, this is a nonjudgmental activity, intended to be a quantitative measure (or "benchmark") that indicates where you currently are. That way, as you grow, change, and try strategies to see what works for you, you have a current assessment against which to measure your forward progress (do you sense the pattern of assessment in this book yet?). Rather than relying on "Well, I guess I'm adding in more exercise" or "I'll take time for myself this weekend," you now have it written down in black-and-white. Setting milestones will help you keep track of your progress as you seek to change and grow. And being vocal about it will boost your leadership impact as you model and give space for others to exercise their resilience muscle too.

Completing this Resilience Ethogram will help you step more fully into being a resilient, instinctual leader. Your ethogram will look entirely different from those of your colleagues, as it relates both to what you're instinctively doing and what you need. Use it on your own or pick an accountability partner to compare notes with. Some people perform a second ethogram a month later to reassess; others wait until three months have passed to ensure more growth over time. Some leaders use this tool with their entire team, to pinpoint efficiencies that can help support and reinforce resilience strategies. Other leaders have added the ethogram to their annual reviews, to document their leadership team's growth and attainment of goals.

Regardless of how you choose to use this tool, it is an important one to add to your toolbox. It is the same exact tool I used to study how brown pelicans spend their time, energy, and focus in the wild, as well as a tool I now use with my clients, so I can attest to its accuracy and impact.

I find it quite liberating to think about resilience from the viewpoint of a pelican. Instead of striving to incorporate mindfulness or quiet periods into my day, I remember how the pelican sat still on the water. Instead of analyzing how I need to harness my energy after a difficult discussion or prepare for an upcoming presentation, I reflect on how the pelican prepared to fly or recuperated after diving into the water. Somehow I don't think the pelican is wondering *how* to do these things. I think it just *does*.

You come built with instinct—take action on it.

You come built with limits—incorporate them.

You come built with resilience.

Unbreakable Law #7 Pro Tips

➤ For anyone in any role seeking to lead teams and facilitate change, embracing resilience as instinct in action must become standard operating procedure.

➤ Resilience is not simply "nice to have"—it is purposefully leaning in to, and taking action on, your instinct about what you uniquely need to lead at your highest level.

➤ By considering the characteristics of a resilient leader and then using an ethogram to evaluate and benchmark your own behavioral patterns, you can figure out (and implement) your individualized resilience tactics and strategies.

Behind the Scenes with KiwiE

Dr. Eugenie Clark (1922–2015) is a legend among shark scientists.[2] Growing up in New York, her interest in marine life was piqued during weekly visits to the aquarium, where she wondered about the life and habits of all creatures, especially sharks. As a young woman in the 1940s, she encountered many obstacles in forging a career as an ichthyologist, such as not being allowed on research ships with male scientists and being turned down for jobs out of concern she might leave to start a family. Undeterred, she came to Florida and started a marine laboratory in 1955 that would evolve into what is today Mote Marine Laboratory.

I learned about Dr. Clark as a college student, and her work inspired my decision to accept a job at Mote in their education department. I moved to Florida in awe of working at the organization founded by such a legendary scientist and surrounded by ongoing, groundbreaking research.

Then I met Genie.

Her eyes would light up when she talked with people—of any age—about sharks. Her office door was completely covered in stickers from her travels and adventures on research vessels all over the world. I shared the stage with her as we gave multiple presentations about exciting developments in Mote's research. She worked alongside me, both of us motivated by how we could continue to refine and communicate our message. She did yoga next to me after work in our conference room. She came to my wedding. My daughter sold her Girl Scout cookies. She always wanted to know what my son was up to.

In a world of expectations, pressures, and achievements, she was as real, accessible, and fun as anyone I've ever met. Even at the age of ninety, she was inviting college students on research trips with her around the world, as she was still fascinated with discovering, teaching, and publishing her findings about the sea.

It is no small thing to say that Dr. Eugenie Clark changed the world and forged a path for women scientists who would come after her. But I will forever treasure all the moments I simply knew her as Genie.

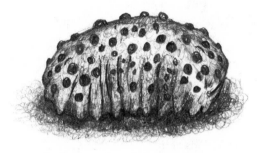

YOU ARE WIRED NOT JUST TO SURVIVE, BUT TO THRIVE

(Take It from Here, Sea Cucumbers)

There is just something about sea cucumbers. I find them fascinating. I can be snorkeling along, admiring the crowds of colorful fish surrounding me, smiling through my mask at the nurse shark resting on the sand and a crab scuttling by, but then my eye will be instantly drawn to the sausage-shaped animal lying motionless on the ocean floor. I might even take a deep breath and dive down to take a closer look. Because this soft-bodied lump of a creature—which to most people may look more like a vegetable than an animal—employs a plethora of unusual adaptations that ensure it's perfectly suited to life on the seafloor.

In my work with aquariums, nature centers, and zoos, I've always loved being a part of the team challenged with designing habitats to suit different animals, discovering what it takes to allow for keeper access, veterinary procedures, enrichment for the animals, visitor engagement, and more.

Designing touch tanks, for example, comes with its own strategies and obstacles, particularly when thinking about accessibility—how to structure an exhibit and a visitor experience so *all* people have the chance to touch animals they may never encounter in the wild.

Bamboo sharks, horseshoe crabs, queen conchs, sea stars, hermit crabs, and yes, sea cucumbers are examples of animals you might find in a touch tank. Some of these—sharks and crabs—are highly mobile and will move around the exhibit on their own. But some—conchs, sea stars, and sea cucumbers—are more sedentary in nature and thus could be beyond a visitor's reach for long periods of time.

To address the challenge of designing our exhibit to ensure all people could experience it equally, we sought to gather valuable input from a wide array of people—especially those who use a wheelchair, which could potentially hinder a complete touch-tank experience. The solution we came up with was to shape one section of the tank like an hourglass so the wheelchairs could fully roll under and people could easily reach in. Watching the exhibit's first visitor using a wheelchair reach over the side of the tank to do something she had never done before—touch a sea cucumber— brought tears to my eyes.

I wanted to start a conversation with her immediately about how amazing sea cucumbers are. I wanted to tell her all about how sea cucumbers may be one of the most unassuming, underestimated creatures on the planet. I wanted to show her a sea star in the tank next to the sea cucumber and have her guess whether she thought these two species were related (they are!). And I wanted to explain that although sea cucumbers are long, squishy, and narrow, and may be firmly in the category (with tube worms and nudibranchs) of creatures-that-don't-look-capable-of-much, that would be a completely wrong assumption. Sea cucumbers can ward off predators much bigger and far more threatening than they are.

But how? she might ask.

Then I could tell her how, when confronted by a predator, sea cucumbers use their first wave of defense: They eject their guts. The fancy scientific term for this is "evisceration." Imagine swimming over to eat a yummy meal and instead finding yourself surrounded by intestines. *Ugh.*

That should be enough to scare off any predator! Then the sea cucumber can simply suck back in its guts and go about its day.

But if for some reason that's not gross enough for a predator, and it decides to enjoy the guts themselves as a snack, the sea cucumber can then use its second line of protection: It simply grows its own guts back.[1]

Yep—it can *regrow its own guts.*

Talk about bouncing back and recovering from difficulties.

That would then inspire me to tell her how I think sea cucumbers are the epitome of resilience. About why I think sea cucumbers provide the perfect animal analogy to showcase how leaders can adapt to life-changing (or even life-threatening) situations and emerge not only alive but stronger than ever. I could follow that with all of my amazing sea-cucumbers-are-the-perfect-resilient-leader examples . . .

But I won't.

For now, I'll just step back and let her experience all the feelings and wonder that go along with touching an animal for the first time.

But if she makes eye contact with me . . .

Gut Resilience Is Your Superpower

There is just something about leaders. I find them fascinating. Often I'll be watching a movie or reading a business leader's words in a news story but wonder more about what went into the hero's decision than about the story's outcome and try to figure out what approach they'll take on an impending disaster or a new opportunity for the company. At meetings and conferences, I take a deep breath and walk right up to (seemingly) intimidating leaders I've never met before, seeking a chance to pick their brain. Because I want to know more about how they, like the sea cucumbers, are perfectly adapted to the situation, their team, the company, or the community—and how, if the surroundings change and there is an impending threat, they are wired to survive and thrive.

In a recent podcast, partners Stacey Dietsch and Elizabeth McNally of global consulting firm McKinsey & Company were asked about capability-building in teams, specifically during this pandemic era as all of us are adapting and growing in our new situation. As the saying goes, thanks to COVID-19, companies now innovate five years in five months, and their people, too, must be equipped not only to keep up, but also to excel in this time of enormous change. From the virtual audience came an interesting question: "Do you see companies investing more in capability-building in tech or in soft skills . . . and is it more difficult to engage in capability-building for soft skills than for tech?"[2]

This is not a new question. I've watched companies struggle with it throughout my career, both from the inside when working for them and from the outside when working with them as a consultant. I've faced this conundrum as a leader myself—responsible for a tight budget, with limited staff hours, and facing intense demands to produce and drive business for the bottom line. That bottom line could be conservation outcomes or financial returns—either way, it's still a bottom line with real-life demands that, if not met, carry even more real-life consequences.

The answer to this perennial question, in my experience, usually lands on the side of technical expertise. People have to know how to use the new software or operate an essential piece of equipment. Team members have to know how to handle the screech owl or the Madagascar hissing cockroaches they will be showing to zoo visitors, or how to scuba dive in order to clean the windows of the beluga whales' underwater habitat.

But technical skills alone will not suffice. They are incomplete without soft skills.

What happens when conflict takes over and can't be resolved? How can emotional intelligence training help team members relate to a newly hired person and establish a fruitful professional relationship? If I, as the leader, don't know how to coach my people to success or communicate effectively with my team, I might as well not show up to work today. A lack of soft skills will derail intended progress, every single time.

All this was ringing in my head as I listened to the McKinsey & Company experts' response.

First, they answered as I expected them to—explaining that a leader, especially in this digital age, needs to embody soft skills in tandem with technical competencies. The same is true for those they lead.

Then they added this: "We talk about the importance of understanding not just the skills that are required, but . . . the mindset that underly [sic] that ability to change because what we're really asking people to do is work in a new way and so that requires not just skills, but it requires different experiences, it requires them to shift their mindset."[3]

Well done—perfectly phrased! This comment expertly captures the third circle of the Wisdom from the Wild Leadership Circles we started with in this book's Introduction. To lead *change* you must develop a quantitative, transparent process (Part 1), which then requires a well-functioning *team* (Part 2), and you must ultimately employ tactics and strategies for *resilience* (Part 3). And a critical element of resilience? Mindset.

Every leader has been confronted with a change they did not think they could (or would) survive, professionally speaking. Every person has faced trepidation over a challenge they thought they should walk away from. You may recognize it from its physical manifestations: dry mouth, sweaty armpits, feeling sick to your stomach. At such moments, leaders of all ages, shapes, and sizes adopt a determined mindset . . . and plow straight ahead.

These leaders stand tall and lock horns with change. They may not fully understand what to do, but they can proactively choose to rely on gut instinct and trust that they will make it through. In school, children learn how to do this by adopting a *growth mindset*. In professional circles, we expand this soft-skills concept using the term "resilience."

The principle of resilience involves a mindset that approaches challenges and hardship with confidence in one's skills and one's ability to persevere, learn, and grow. Yet although we strive for this mindset, we can all be guilty of overthinking things and getting in our own way. Facing unprecedented stress, we begin to flounder.

This is when we need to channel our inner sea cucumber.

How do sea cucumbers unleash their resilience in times of great stress? By simply doing as they were born to do—by using (quite literally, in their case) their guts.

In people, doing "what comes naturally" can mean either an innate or a learned skill, depending on the stress that you encounter. Have you ever been told, "Wow, I don't know how you did it!" and responded with "Well, I don't know, either. I just kept putting one foot in front of the other, because what choice did I have?"

Then you have tapped into your gut resilience—your inner sea cucumber.

Adapt to Challenging Situations by Channeling Your Inner Sea Cucumber

My career has taken twists and turns I never could have foreseen. When I first toured behind the scenes at an aquarium, I had no idea that one day I would lead those tours myself. When I first put my hand up and volunteered to attend a conference, I had no idea that years later I would speak at conferences in places like Victoria, British Columbia; Cornwall, England; and Tortola, British Virgin Islands. And when I boarded a roller coaster at my favorite zoo and amusement park, I had no idea that someday I would be an integral part of opening the largest expansion to date in the company's history.

That experience, of being part of a team opening an exhibit featuring tigers and orangutans, was a far cry from my time spent thinking about how to highlight animals a little lower on the activity scale—sea cucumbers.

Sea cucumbers barely seem to move at all. Watching them very closely, however, you may be able to identify circular mounds of sand next to one end of the animal—sea cucumber poop piles! That's your only clue as to which end is the "head" and which is the "rear." As the sea cucumber inches across the ocean floor, it sucks in sand through its mouth, filtering and digesting all the yummy microscopic critters hiding in the sand, and then poops out the remaining sand. The animal itself may not even seem all that energetic—until it must quickly react to the challenging situation of potentially being eaten.

The sea cucumber's line of defense—evisceration—can be a success-ful solution to saving its life.[4] But on the off chance the sea cucumber's ejected guts look quite tasty themselves, and the predator decides to devour them instead? Both animals will still go on to live another day, albeit sans intestines for the sea cucumber, as it regrows its guts over the next few weeks.

Bottom line: In a not just seemingly challenging but actually life-threatening situation, when a sea cucumber literally loses internal organs, it will survive.

What could be more resilient than that?

It is wired to survive, and even thrive, in the most harrowing of circumstances—to face a threat to its very existence—and then live to see another day and continue to play its important role in the oceanic ecosystem.

We need to do this as leaders.

Before I knew how to channel my inner sea cucumber, I learned about a prestigious club at my university that everyone wanted to join. It was extremely active on campus, working to help students succeed at every level, and was known to be a breeding ground for professional competencies. I watched as members of this club graduated with cultivated leadership skills and got hired for sought-after jobs. So, I filled out the application and was granted an interview.

I arrived, nervously sat down, and faced the two club members on the other side of the table. They smiled encouragingly, and all was going well … until they asked me: "Tell us about three strengths and three weaknesses that you have."

I was stunned into silence, trying to figure out a way around this thorny question. Answering honestly seemed like a very foolish idea. The interviewers waited patiently as I mulled over possible answers that wouldn't make me look either too conceited or downright unqualified, but that offered just enough honesty and transparency to paint me in a good-enough light. I felt stuck between options, put on the spot. I just couldn't think of anything to say. I sat silent.

The panic started to arise in my body. Sweat formed on my brow, and the feeling of queasiness took over my belly. The awkward silence extended and eventually made even the interviewers seem uncomfortable.

"Perhaps there's a scenario you could tell us about in which you helped someone out who really needed it?" they prodded. "Maybe a case in which you were out of your depth but rose to the occasion? What can you tell us to help us get to know you better?"

I remained silent. The moment sat frozen in time.

I never did rise to the challenge, never did think of a "good enough" answer, so the interviewers and I stumbled along to the next question and on to the end of the interview.

I didn't get invited to join that prestigious club.

I have never let that happen to me in an interview again. I have arrived overprepared with practiced answers to a host of questions. I have learned techniques to calm my nerves and stall for time while creating an answer in my head that will satisfy the interviewer while steering the conversation toward my strengths. I have prepared myself to choose words carefully, phrasing them with measured tone and even cadence, and to present an aura of professional presence.

Even when faced with the toughest interviewer in my career, at a world-renowned global company for a highly sought-after position—during which the recruiter fired off questions nonstop and never once broke eye contact with me for thirty long minutes—I took deep breaths from my belly to calm that just-under-the-surface feeling of queasiness and managed to project a cool demeanor.

I rose to the challenge.

Cool as a sea cucumber.

Flex Your Resilience Muscle and Emerge Stronger than Ever

The ability to rely on your inner sea cucumber, to flex this muscle of resilience, is innate for some people, even for those who are unassuming or underestimated by those around them. These people may not project the stereotypical leadership skills associated with the job at hand. But just as animals don't always behave the way we expect them to behave, people don't always pay attention to the expectations of others. They surprise you with new ways of doing things. They aren't limited by the opinions of others.

Like the sea cucumber, they just are who they are.

My son, Kepler, came out of the womb calm, cool, and collected. He could hold his head up well before he was expected to do so, and he showed empathy for others at a very young age. I often wouldn't know where he was in the house except for when he would call out, "I'm OK!" Whether staying at home for the day or headed to Disney World, he's always "fine" with whatever we're doing. He is inquisitive, witty, and highly imaginative—every drawing, object, or story eventually becomes about robots. I am immensely proud to be his mother.

Over the years, few people have outwardly remarked on my son's leadership skills, as compared to my daughter's. Maybe it's because he's quiet at first or busy creating stories in his head instead of engaging in someone else's moment. Yet Kepler, like his sister, is a born leader.

One day at a first-grade classmate's birthday party, two mothers approached me excitedly and said, "We're so thankful for Kepler! We tell our kids to follow his lead so they can deal with the class bully, too."

I nodded, smiling along with them.

Wait a minute . . . there's a class bully?

I attempted to conceal my shock as they went on to explain that the bully would go up to kids at recess and terrorize each one before moving on to the next. But when the bully approached my son, instead of backing away or even looking nervous, Kepler stared the bully down and yelled, in his loudest voice, "*Ahhhhhh!*"

Confused by and unsure how to handle the situation, the bully backed away and never bothered Kepler again.

My son employed a quiet, understated, subtle trait of leadership—a profound reliance on his ability to stand up for himself—far younger than I had been when I managed to find inner resilience in the face of such stress. He didn't question, didn't overthink, didn't even stop to consider that he couldn't do it. When the bully approached him, possibly assuming that Kepler's naturally reserved nature made him an easy mark, Kepler stood his ground. He just did. He just was.

In a threatening situation, he channeled his inner sea cucumber, faced down adversity, and emerged with a sense of his own power. The bully retreated. And Kepler went about his day.

Never underestimate a sea cucumber.

As leaders, we inevitably face times when we feel overwhelmed. Sometimes it seems that the pathway forward may swallow us whole. Sometimes you feel lost or in way over your head. You may be fearful of losing your position, failing at an important project, or confronting any of the thousands of realities that leaders must confront over time. And then there are all the other responsibilities you are balancing: taking classes, keeping up with certifications, creating quality family time, getting exercise, serving on nonprofit boards, volunteering your time, and more. Yet we become conditioned to this continual pressure—to needing to adjust, figuring out how to adapt, feeling as if we must keep moving in order to lead.

Sometimes we bully ourselves into maintaining this overcommitted life, until we realize it's time to put our foot down. Sometimes we have to say no—our own form of sea cucumber evisceration. Guess what will happen if we do listen to our gut and reject something that feeds the hungry predator of our all-consuming ambition? It will leave us alone. It will go away and let us be. And we can go about our day (or our life).

And guess what else? As hard as it can be to say no, the truth is: You will survive, and even thrive. You may even turn out stronger and more resilient than before.

You can say no and still live to see another day.

I have mentors and role models who never say no. I've watched them work and achieve and have incredible impact on their teams and their mission. I've seen them show up early, stay late, attend events every day of the week, and sign up for more. Their way of life was once exciting for me and one that I aspired to.

Then life moved on, and I've watched these same individuals battle health issues and struggle to balance their personal and professional obligations. I've seen them grow tired, overworked, overstressed, and stretched too thin. Some of them have stood up and said no, when I didn't think they ever would. Some have continued to say yes long beyond the limits of their resilience.

At each stage of leadership and life, a different type of resilience may be required. As an experienced leader, I have learned to flex my resilience muscle by saying no.

Sometimes it feels better than other times. Sometimes I can barely stomach it. The larger and more exciting the opportunity, the more intense the feeling of wanting to say yes. My comfort comes in knowing that saying no makes room for someone else to benefit from the opportunity, fill the position, or impact the mission, community, or bottom line.

Saying no still gives me a queasy feeling in my belly, but it also means saying yes to the next thing. I don't know what that thing will be, but I'll be ready. By making space in my life to strengthen other areas, I can ensure that I'm not just surviving, but thriving.

And each time, after each tough challenge, I emerge stronger in some way.

Can you? Yes.

Because you are wired not just to *survive*, but to *thrive*.

Will you? Only if you choose to channel your inner sea cucumber.

Unbreakable Law #8 Pro Tips

➤ In times of great uncertainty, when you may feel that success is out of reach, channeling your inner sea cucumber will help you find gut-level resilience and remember that you are biologically designed to thrive.

➤ Like the sea cucumber, you are perfectly designed to be what you are, with impressive points of leverage: the ability to adapt in challenging situations and the power to ward off threats and start anew.

➤ Whether people expect too much from you or not enough, by flexing your resilience muscle (and saying no when necessary) you can survive another day and even emerge stronger, leading as you were designed to lead.

Behind the Scenes with KiwiE

I answered the phone one day at the aquarium, and the woman on the other end of the line was ecstatic but confused. Her words came rushing out as she described how she and her children found a sea turtle and were concerned that it was not in the ocean. So, they took it to the beach and put it into the water, waited, and after a few minutes, it crawled back out. Then they would take it back into the waves, wait, and sure enough it kept crawling back out . . .

She was so distraught, trying to help this lost turtle, and she couldn't figure out what was going wrong.

I took a big breath before responding.

"Thank you so much for calling us, and I understand your concern to help this animal," I began. "Can I ask—does it happen to have legs with webbed feet? Or does it have legs and feet that look more like flippers?"

She asked me to hang on, and she put the phone down, then came back on the line.

"It has legs with webbed feet!" she exclaimed.

I took another big breath and replied, "You've made an excellent observation. And, again, I'm so thankful that you and your kids are looking out for wildlife. But, I think, since it has legs with feet and not flippers, that you have found a freshwater turtle, not a sea turtle. This animal would be far happier living near a pond than in the ocean. The next time it crawls out of the sea, why don't you take it to a pond or lake near where you found it?"

"Ah, OK. We'll do that," she replied.

EVEN CHEETAHS SLOW DOWN

(*I'm Looking at You, Leaders*)

I've always been a cat person. I grew up with Charcoal, a black cat that we took home from my grandparents' barn, whose hunting prowess was on par with a jaguar's. As a young adult, I inherited a 90-gallon tank and went to the pet store thinking I'd come home with fish—but instead adopted Tigerlily and Angel, two shelter cats who became fabulous wing-women as I embarked on single life in the big city.

Twenty-odd years later, I found myself at the local shelter once again, searching for the right match. I scanned the cats roaming free in each room until he caught my eye: a tall, regal-looking, eight-year-old male with prominent markings and a calm demeanor, waiting patiently and confidently by the door for his soon-to-be family to walk in.

We named him Max.

A loyal companion ever since, he carries himself like an African cat called a serval. Servals have long legs and muscular, slender bodies, traits that help them hide in tall grasses and expertly hunt down prey. They can jump more than nine feet straight up in the air to catch a bird—a trait I see when Max leaps into the air himself.

But when I hear his high-pitched meow from the other room, he sounds just like a cheetah.

What? you might be asking. *A big cat like a cheetah—I thought it would roar!*

Nope—a cheetah meows and purrs, just like my housecat, Max.

Cheetahs and Max have something else in common: their innate sense to know when to rest. For the African speedsters, this happens after maxing out at roughly 60 miles per hour as they chase their prey for about thirty seconds. For my cozy house cat, this happens most of the day.

This ability to take action when it's time to relax, take a break, and avoid wasted effort can induce jealousy in those of us who consider ourselves a type A personality. While we might wonder, ponder, consider, plan, and attempt to schedule in a time to rest, cheetahs (and Max) don't overthink—they just do.

They slow down.

Your Team Needs You to Be a Cheetah

Whether you like it or not, and whether you admit it or not, you will eventually slow down. And that's OK. In fact, that's what's designed to happen.

This is a biological fact. *All* leaders, like *all* cheetahs, slow down at some point, whether by choice or by consequence. The fastest land animal in the world cannot run at top speed forever.[1] Even this marathon runner must slow down at times to rest and recover—and so must you.

So why do we, as leaders, assume we can run at top speed forever?

To be fair, maybe we don't really assume we can keep this speed up forever. Maybe it's that we just feel compelled to keep going: spinning all the plates, juggling all the balls, stretching ourselves so thin that it's impossible not to break. Maybe we arrive at a point of constant movement and it dulls us into a sense of normalcy, into thinking that top (and continual) speed is the only speed. Our identity can even become impossibly intertwined with being busy, leading, doing, bettering ourselves, challenging the status quo, creating, mattering.

I would tell you to just *stop*. But that's not easy, and nor is it reality. It's not what you want to do, either! You want to lead. You *need* to lead. Like the cheetah wants to—needs to—run.

But guess what? Both you and the cheetah need to rest and recover before running again. And to truly lead the way your team needs to you to lead? They need you to embrace this lesson—this biological certainty—from the cheetah. They need you to intentionally determine when and how to do this as it makes sense for you, instead of falling prey to the sudden stop that will be inflicted upon you by forces out of your control. They need you to, at times, *slow down*.

Of all the Unbreakable Laws of leadership in this book, this one is the guiding force because I believe it is the very essence of being a leader. The cheetah's need to preserve its energy for the challenges ahead is the very definition of an Unbreakable Law of nature—a fact that you cannot argue with and cannot manipulate. It's a fundamental, biological reality that is equally inevitable in the animal kingdom and the business world.

I recognize that there are many possible motivations behind why you as a leader—like me—run and run and run and run.

Maybe you run because you are a natural leader of people. You want them to grow, work together as team, and succeed. You want them to achieve goals, feel valued, and effect change. And you want the same things for yourself. So when you see that next opportunity—or maybe a threat—coming toward you, you put all your effort into leadership. You run.

Maybe you run out of sheer excitement. You can't get an idea out of your head until you see it in action. When you see the possibilities on the horizon, you know if you just keep going, just a little bit longer, you will be there. You can make it happen. So, you run.

Maybe you run because that's what you see other leaders do. The people around you—colleagues, mentors, friends, family members—are also leaders, all attempting to run at top speed for as long as they can. You buy into this version of leadership. So, you keep running too.

No matter your motivation now, and no matter how it may change over time, as a leader, you run as fast as you can. You run toward the goal, the competition, the idea, the threat, the innovation, the challenge, the possibility. You run knowing you do so for, alongside, behind, and with your people.

We've all seen those images of a cheetah running. Do an internet search for "cheetah," and the first images you'll most likely see are those of a cheetah running, hunting, or staring down its objective. By far the cheetah's most commonly discussed characteristic is its speed—not how much it needs to rest after accelerating so quickly. But to miss out on discussing the next part, the recovery, is to ignore the other side of the coin.

This truth—no resting means no running—is one that we, as leaders, may not want to embrace.

But the cheetahs do.

Saying Yes to Everything Is the Quickest Way to Slow Down

What does this mean for you as a leader? You need to slow down *proactively* before your mind, body, and spirit do so *reactively*.

Like many people in a leadership position, I have a habit of saying yes to everything. All of us have our own reasons for this, and exploring your personal motivation for trying to run at top speed, for as long as you are physically, mentally, and emotionally able, is key to understanding how you can choose to stop, rest, and recharge.

Many leaders do not know why they can't say no.

But I do.

It's because I look way too young for my age, which has led to people assuming I don't know what I'm doing, I can't possibly be in charge, or that I simply don't belong or deserve to be there.

Which has led to decades of working harder—not smarter—to prove them wrong. As far as I know, this has done little to shift the needle of *their* opinions, but it has resulted in extensive damage to *my* energy, health, identity, and psyche.

To be born with a deep, insatiable drive to innovate and achieve results is a blessing—although not necessarily when combined with the genes that make you look younger than you really are. On the surface, it seems like a winning combination. I don't shy away from hard work, and I'm intrinsically motivated, with little need for external recognition on the job—I know when I've dialed it in, done my due diligence, and truly considered each outcome. And I am filled with gratitude for my grandmother's genes and the lack of expensive anti-aging creams in my bathroom cupboard.

Except when it comes to my work and perceived influence, where looking younger has not served me well.

I have interviewed for jobs while continually feeling the need to spotlight my expertise and formal education because the questions were tilted toward my perceived inexperience.

I have been assumed to be the intern—when in fact, I was the boss, four hierarchical levels up the ladder.

I have been passed over during conversations by people who thought I didn't have anything to contribute—when in fact, we were peers.

I have been denied consulting opportunities and given the explanation that I didn't have much to offer as I was new to the field—when in fact, I brought decades of experience to the project.

I have arrived to deliver a keynote address and been asked by the staff whether I'd registered for the event, which was targeted to young, early-career professionals.

All this may seem flattering, but I have not experienced it that way. On the contrary, I have spent my whole career resenting it and overcompensating.

These seemingly innocent interactions have negatively impacted my self-esteem and led to a feeling of imposter syndrome. I've felt I had to continually *prove* that I belonged, that I *knew* what to do, that I had *earned* my seat at the table. The weight of proof was time-consuming, heavy, and cumbersome. But I believed I had something to contribute and could positively affect both the projects and the bottom line. I wasn't ready to accept someone else's limiting opinion but also didn't know how to permanently change it. So I did the only thing I could think of.

I dug in and worked harder.

I spent longer hours on the job and was rewarded for this as a young professional. So, I worked even harder.

I chose professional attire suitable for someone decades older than me, with no trace of my individuality, hoping to lend gravitas to my public persona and not look like the youngest person in the room.

I joined more boards, volunteered for more committees, found more volunteer work, attended more trainings, earned more certifications.

I led with my business card in networking, making sure people knew my title, and forgot to care whether they knew my work or even knew me as a person.

I found myself climbing all the corporate ladders, earning all the job titles and accolades I could: managing a team, filling two leadership roles simultaneously, teaching kickboxing and Pilates, tutoring students in math and science, and attempting to maintain a twentysomething's social life.

In the middle of all of this consistent, over-the-top effort, I applied for a yearlong graduate program overseas.

And a few short months later, I was on a plane to New Zealand.

Flying away from my established life, I was feeling nervous, excited, unsure—but most of all, I was feeling so, so tired. I was tired of climbing the corporate ladder . . . of trying to prove that I mattered . . . of working to be needed . . . of being a champion for every cause in the world . . . of being a leader of people . . . of making sure my team was taken care of . . . of not remembering how or when to take care of myself.

When I flew away from the United States, I left everything I knew and everything familiar in the way I *lived*: up early in the morning, work all day, drive to night school or my second job tutoring or the gym, return home to answer email, attempt to keep up with relationships, fall into bed dead asleep, and get up to do it all again the next day.

Now after a sixteen-hour flight over the Pacific Ocean, it was like I had stepped out of a vortex. I found myself living in a small flat at the top of a hill, with three TV channels, no internet, and no car. Walking to graduate classes, strolling through the botanic garden on the way home, enjoying a book, meeting new friends, joining a local photography club, researching places to explore . . . that year of my life was magical.

And with no potential social media distractions (as this predated them all!) I was not tempted to post or share my experiences.

I slowed down and simply lived.

Although I didn't fully realize it, I had desperately needed a break. I was beyond exhausted—quite literally used up. I barely recognized myself anymore.

Moving to the other side of the world offered me a vital chance to recharge and reset how I approached my life and my work. It was an opportunity to remember what I actually cared about—why I did what I did—and maybe figure out how *not* to prove myself anymore. It was a chance to just be.

And frankly, I was a little miffed.

I had grown so accustomed to working hard in every aspect of my life that I was totally caught off guard with how I was feeling. Had my work somehow betrayed me? Had my efforts to work harder and smarter, to prove I was "supposed to" be there, somehow backfired?

I had given so much of my energy and my life to one thing—my work—and that's what I had become. Suddenly, with no high-powered job, no team to work with and work for, no purpose bigger than myself, no constant need to prove myself, I was in uncharted waters.

I don't remember the moment my work became my identity. But I do remember the agonizing transition as I was jarred unwillingly into undoing this reality.

There's a typical conversation we encounter thousands of times, a casual interaction commonplace in the United States whether you're at an official professional networking event or out for a family picnic. It's the question we've all answered throughout our lives:

So what do you do?

And I had my typical answer all set to go.

Well, I'm a vice president at X organization, and I'm living in New Zealand for a year on a Rotary Ambassadorial scholarship studying in the graduate school at the University of Otago.

Then I planned to field the typical follow-up questions about my job, my studies, my goals—even expecting the typical comment I was used to hearing, *But you look so young!*

Remember, I was used to defining myself as a leader—that someone else had hired, with a title someone else had given me, from an organization that someone else had started. And I *was* a leader, but every other aspect of my identity had come from outside of myself, involving decisions not entirely within my control.

But that initial question never came.

People in New Zealand were not expressly interested in what I did for a living. They didn't seem to think about how old or young I looked. They were not even familiar with the organization I worked for in the U.S. They seemed only interested in connecting and having an actual conversation.

They wanted to get to know *me.*

What kinds of things was I interested in? What trails had I explored in New Zealand? What was life like in New Zealand compared with life in the States? What was I planning to discover next outdoors?

Yet, at each networking event, I built up my anticipation.

Surely today, I thought, *at this social, with these professionals, someone will inquire about the job I hold.*

Nope.

But surely tonight, at this dinner, someone will ask about the work I've dedicated my life to.

Not happening.

Well, this weekend, someone will wonder about the changes I want to make and the legacy I want to leave with my career.

No go.

Each person I met continued to ask about, wonder about, and care about me—just me. They didn't care if I looked "too young to be thirty" or that I was on my second graduate degree. They weren't all that impressed that I was a vice president. They wanted to know about me—as a human, a unique individual.

This ground me to a halt. How would I respond to these new questions? What was I going to ask *them* to get the conversation going? Gone was the fast-paced but familiar box I lived in, by which I defined myself and measured whether I was succeeding. Slowly disappearing was my desire to prove I was worthy or up to the challenge.

Instead, I found myself feeling *safe, but not comfortable.* I began to ask myself new questions: What does leadership look like outside of an outwardly defined role? What do I really value, in my finite life, regarding the impact I can make? How can I harness this awareness to move away from proving myself, and into working toward and within my purpose?

When I returned from my year in New Zealand, I was changed in both tangible and intangible ways. I had lived a whole different life, and I knew I would never be the same. Not as a daughter, not as a wife and mother, not as a friend, and certainly not as a leader. Even though my brain didn't know it consciously, every other part of me knew I needed a rest. And I didn't just need time to wind down.

It was time to truly rest in the shade, alongside the cheetahs.

Care for Yourself Today, So You Can Lead Others Tomorrow

Fifteen years after I returned from New Zealand, I was working with a client to design a training program: gathering specific input, researching the organization's current needs, customizing the program, delivering the training, and following up with individual coaching sessions to help each person apply tactics to their specific role and responsibilities.

One person in particular caught my attention during this process, as he was a sponge for new ideas and remarkably responsive to specific constructive feedback. He worked hard and was a valuable, high-performing member of his team—labeled a "high potential" in the organization's succession planning.

He also happened to be young and new to this particular company, recently hired out of college and working in a field in which he had limited technical knowledge. But he was eager and enthusiastic, an active participant and a joy to have in class. He showed up first and left last. He implemented new ideas and always wanted to know more.

I recognized myself in him immediately.

During our one-on-one coaching session, his easy-going yet driven demeanor quickly gave way to rampant insecurity and fear. He explained the root of his uncertainty: extreme self-consciousness about being the youngest person in every room. He felt out of his depth and wanted to prove himself, so he continued to work harder than most and push through exhaustion to anticipate questions, develop new ideas, and make sure he was needed and valued.

I was talking to me fifteen years ago.

The continual cycle of running and running and running to prove himself resounded loudly in the recesses of my brain. I recalled that feeling as if it were yesterday, not years ago.

Yet I could tell that he had not yet translated the unique journey of his miles of experience and moments of decision-making into real-time leadership prowess. He had not realized the fundamental role resilience plays in showing up as your best self to lead. Instead, he was running and running and running with no end in sight.

It was my turn to tell him what I wish someone had told me all those years ago.

I advised him to respect the learning curve, adjust where possible, and even learn new ways to connect with people—strategies that would *work for him* so he wasn't working himself to the bone.

I reminded him that no matter what, someday it would be the opposite: He would be surrounded by people new to the industry bringing fresh, unbridled ideas. The world is always growing and evolving, and his life would have different phases. There isn't anything special about being "youngest" or "oldest." He would always need to adjust his strategies accordingly and to harness the power of resilience.

I could fill pages and pages with ideas on self-care: strategies you can employ to strive for a more realistic work-life balance, the importance of mental and emotional health in direct correlation to your physical health, why slowing down is so essential to your ability to lead change that will stick. All these ideas are valid and needed in the leadership development space. But to take care of yourself in this way, you have to be willing to slow down in the first place.

For leaders, the danger is not some predator lurking around the corner or some prey that might outrun you. The danger is not the other cheetahs that are naturally faster or may outcompete you for resources. The danger is not outside of you. The true danger that can keep you from leading how you were born to lead—from driving change, leading teams, and harnessing resilience—is within you.

When you become so intertwined with your professional goals, your work, your responsibilities to your team—with needing to work harder and harder and harder—your work becomes your identity. You lose the ability to differentiate between the purpose of your work and the purpose of your being. You can't step away, can't take a break, can't slow down.

Your work *becomes* you.

Unless you want to be a leader who is only half-focused on the moment in front of you, distracted by mental exhaustion and emotional fatigue, the necessary pivot you need to make at this point to flex your resilience muscle has three components:

Lean in to the *miles* of experience you're bringing to your particular role and the reassurance that you are exactly where you need to be. Lean in to who you are now.

Trust that, in this *moment*, you know the best way to approach the situation at hand, building off of your unique expertise and perspective. Trust that you are here for a reason and work intently focused on that purpose.

And take action on your need to *rest and recharge* in ways that make sense to you, so you can show up the next day and do it all over again. Intentionally step outside yourself (or maybe even outdoors!) for a moment and recharge who you are.

It may not feel necessary or important now, but *you must slow down from time to time in order to be successful.* Trust me—I know what happens when you don't. I know how it feels when you wake up on the other side of the world, not knowing fully who you are and what you stand for. Don't go too far down that path.

Sometimes it may get so loud in your own head that it's hard to hear your instinctual voice. Life gets busy, and leadership is demanding. You may suddenly find yourself behind the eight ball, breaking down before you've had a chance to recharge.

These three components will help you stop yourself from getting up and sprinting ahead before you're ready, and they can inform how you rest in the shade too. This is the way you lead at your highest level—by listening to who you are, from the inside out.

Cheetahs, like leaders, have their unique abilities and niche in the world. They fill a specific role in the wild like no other animal can. The health and vitality of wildlife and wild places needs them to just be cheetahs—no more, no less.

And oh, how cheetahs are built for their niche in nature.

The huge leg muscles? Perfectly tailored to propel the cheetah at top speed. For you as a leader, your powerful momentum is built up over years of experience—in both good times and bad—which makes you strong enough to lead at top speed.

The small, lightweight body? Perfectly designed to support the cheetah's quick movements and aerodynamically built for speed. For you as a leader, your light, lean frame is honed over the expertise gained through countless training programs, educational seminars, and mentor programs during which you proactively decided which strategies worked for you and left the rest behind.

The loose hips and shoulder joints? They allow a cheetah to change direction at a moment's notice, lengthen its stride and accelerate. As a leader, you gained that agility by learning to stay focused on the goal ahead, hypervigilant of the need or opportunity to change direction and stay on course.

The flexible spine? Expertly capable of operating like a spring, providing the base from which its legs can propel forward at full throttle. For you as a leader, that flexibility holds you to the core of your identity in tandem with pivoting to allow for various perspectives and valuable insight provided by your team.

The semi-retractable claws? Poised to help the cheetah dig in for speed when it needs to. For you as a leader, they represent the tenacity and determination you need to dig in for the long haul when you need to and then to pull back when it's time.

Ultimately, like the cheetah, you are driven by the desire—and the need—to run.

And ultimately, like the cheetah, you must rest to be truly resilient, so you can do it all again the next day.

You need to slow down so you can lead how you were born to lead.

Unbreakable Law #9 Pro Tips

➤ Your team, your company, your community, and your purpose need you to show up as a cheetah, ready to run at top speed—and willing to accept that slowing down is both inevitable and preferable.

➤ Saying yes to everything or seeking to live up to others' perception of you is the quickest way to discover that, whether voluntarily or involuntarily, you will someday reach a point where you cannot keep going at that pace anymore.

➤ To be the best leader you can be, you need to be a cheetah in top form, which means flexing your resilience muscle and trusting that you are perfectly designed to do so.

Resilience is instinct in action.
You are wired not just to survive, but to thrive.

And even cheetahs slow down.

You are built for resilience.
You are wired to listen to your instinct and thrive in the face of adversity.
Rest in the shade so you can come back even faster than before.

Behind the Scenes with KiwiE

As part of a multi-day outdoor leadership retreat, our group found ourselves exhausted at the end of the first morning of team-building. We'd been collaborating, communicating, and co-creating so many creative solutions and new methods of working together that we were ready for a break.

Our leader led us on a short hike to a tall, broad, gorgeous tree. He turned with an excited smile (and mischievous gleam in his eye that I would later recall) and told us we were free to climb this tree! Into the tree we went, twenty people spreading out onto limbs and branches to relax in the shade.

After a few peaceful moments, I noticed that items were now being passed up from the ground to various members of our group. I saw a loaf of bread pass by, then a jar of mayonnaise, followed by a package of sliced ham, a head of lettuce, knives, plates, napkins, and more. I played along until abruptly the motion stopped and I found myself holding a package of sliced cheese. Then I heard the voice of our leader calling out from the base of the tree:

"OK—lunch time!"

We started glancing around at each other as it slowly dawned on us. We were having lunch—hooray!—in a tree . . .?

Without hesitation, a team member to my right called out, "I've got the bread! Who has the plates?"

Another person from the far side of the tree answered, and we began passing condiments, utensils, and food back and forth, all the while balancing somewhat precariously in the arms of a tree, several feet off the ground.

(*continued*)

The result? A well-earned lunch, only made possible with the unprecedented collaboration of people in a tree. And arguably the best, most memorable team-building experience I have ever been a part of.

INSTINCTUAL LEADERSHIP FIELD GUIDE: RESILIENCE

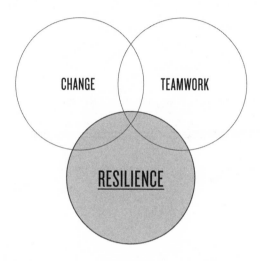

Now it's your turn! I invite you to reflect on the ideas, tools, case studies, and insights provided in the previous three chapters, all centered on the leadership concept of *resilience*. The following prompts will guide your reflections as you critically evaluate the relevant lessons and how they can best help you.

To target your answers and make them as immediately applicable as possible, I advise you to think about your answers only as related to your current role as a leader.

As you're reflecting, think of your own examples, case studies, and maybe even a few animal stories! Jot those down in the white spaces around these prompts. They will come in handy as you refer back to this book and pass along these lessons to others in your life.

This is how you truly personalize and implement the wisdom from the wild as it relates to resilience.

1. The idea that resonated with me the most from Part 3 is:

_____

_____

_____

_____

2. The biggest opportunity regarding resilience I can see is:

_____

_____

_____

_____

3. The biggest challenge regarding resilience I foresee is:

_____

_____

_____

_____

4. I am most excited about:

_____

_____

_____

_____

5. I will take action and hold myself accountable by filling in one or two ideas (no more than that!) in each box below, building off of the ideas in Part 3, to be implemented in the next three months:

STOP	KEEP DOING	START

6. I will pay it forward and help others that I lead and influence by:

_____

_____

_____

_____

CONCLUSION

There is a reason you picked up this book. Maybe you've always been fascinated by animals. Maybe an early visit as a child to your local aquarium or nature center, or even a recent walk outdoors with friends, solidified this fascination.

Maybe you were looking for a fun way to think about challenges or a creative way to talk to your team. Maybe you were in the mood for some inspiration and wanted to feed your leadership soul.

Maybe you have a significant challenge that is difficult to talk about or even put into words. Maybe you were feeling invigorated by a new possibility that no one even knows about yet.

Or maybe you've read every other leadership book out there and want to think about things in a new way.

Regardless of how this book found its way into your hands, I hope this exploration through the nine Unbreakable Laws of leadership from the animal kingdom has unleashed your individual potential to impact the world in the way that only you can. Make no mistake: Leadership skills can be (and are) taught, but they only build on the fundamental qualities that make you the insightful, instinctual leader you already are.

Wildlife and wild places provide us with unique insights and inspiration for your journey as a leader, because the lessons are so deeply rooted in instinct. Mother Nature also has a way of being totally up front and honest about everything, no beating around the bush. You can't fool her—she has a way of peeling back any pretense or unearned swagger you may bring outdoors, while also enveloping you in both challenges and joy. Because the magic of learning from nature is rooted in this: Wildlife and wild places will not ask you to be anything other than yourself and will continue to provide both motivation and stillness as you grow and change over time.

Nature is speaking. Are we listening?

The lessons shared in this book are meant to harness the knowledge we can learn from the animal kingdom.

The case studies and stories are meant to personalize these ideas through experience.

And as you take these thoughts and insights into your own world—testing, re-shaping, customizing, and applying—you truly step into the wisdom that can be learned from the wild.

None of us—not even the animals—begins with wisdom.

As our exploration through the nine Unbreakable Laws of leadership comes to an end, I leave you with one final tool in the form of a diagram for your leadership journey. Draw it into your journal, on a whiteboard, or simply in your mind, as you reflect on and bring forward ideas from this book into your future as a leader.

At the base of the pyramid is *knowledge*, the foundational understanding and skills you possess now and will continue to add to over time. Just as a sea turtle hatchling innately knows to head to the sea and a young polar bear watches its mother to learn how to hunt, leadership is about learning the ropes, finding your way, accepting a new challenge, and learning all over again. Your knowledge is a wide and solid base on which you can continue to build your whole career.

Layered on top of your foundational knowledge are the unparalleled learnings from *experience*. No book, training program, or executive coach can ever truly explain what to do in each situation, and experience is earned as you live and breathe leadership fundamentals and ideas. I've heard scientists say time and time again that on paper, something shouldn't work—the water quality was not what they expected, or the environmental conditions were less than ideal for particular organisms to survive. Yet, the organisms were there. The scientists had learned all they could from research, mentors, and colleagues, but their experience would be a game changer. In fact, this pivotal moment of discovering something new from personal experience would form the basis of their future work and propel them to have a lasting impact as only they could.

And at the pinnacle of the pyramid, as you continually gain experience, you will earn *wisdom*. Leadership prowess is built on the unique miles and moments we bring to each situation, but there is no shortcut to wisdom. It is acquired through boots-on-the-ground experience combined with intentional reflective periods of rest.

This is how your impact and influence as a leader becomes as unique as your fingerprint.

This is how you fully leverage and celebrate the wisdom from the wild.

Wisdom from the Wild Leadership Pyramid

Ultimately, which animals (or plants) described in this book did you connect with as a leader?

Are you like the sea cucumber, who rises up to the next challenge even after great adversity? Are you like the coral, who has felt overlooked but is actually an integral part of the organization's solid foundation? Are you like the naked mole rat, whose unconventional approach to teamwork may seem strange to some but highly effective to others? Are you like the sea turtle, so deeply rooted in your purpose that you will keep going through miles and miles of ups and downs to see it through?

I believe you can learn from all of these creatures, and so many more that I could have written about.

Just as I continue to do.

Understanding that the cheetah, the fastest land animal on earth, will always slow down—no matter what!—makes me feel better in taking the time I need to excel as a leader. Even if a cheetah is not successful in catching its prey, it still slows down. It does not stop to beat itself up about why it wasn't successful. Sure, maybe the cheetah reflects for a moment, with an innate sense of considering a different route in the future. But no matter what, the cheetah instinctively knows that resting and recharging is not just a nice-to-have; it's an essential if the cheetah wants to run at top speed the next time.

When I walk outdoors and see a tree I don't know or wonder why a frog is making a strange call I've never heard, I remember that I don't always have to know all the answers.

And when I see a baby bird peering tentatively over the side of its nest, I wonder what it might be thinking. I wonder if it questions whether or not it should lean in to its instinctual nature. I wonder if the little bird truly believes it can trust its gut and learn to fly.

Like this baby bird will discover, opportunities to lean in to your unique gifts as a leader—to follow your instinctual road and learn to fly!—will always be yours for the taking. But only if you're willing to take a chance, lean over the edge, and trust your gut.

Just as I continue to discover.

Not long ago, I was visiting a friend and former boss who has known me throughout my entire career. Our families were spending the day together and decided to visit a local go-kart track. This was not just any go-kart track, but one that was across the street from the famed Indianapolis 500 racetrack.

My two children suited up for the kids' race. They watched the safety video, conferred with each other, and made sure they knew the rules of the road as they prepared to start their race. They made a deal to look out for each other on the road and not make it a competition between the two of them. Even so, I felt more than a little nervous as they donned their full-face helmets and neck braces. I could barely make out their facial features, but I exhaled into the moment and gave them a big reassuring smile.

Then I glanced up at the TV monitors announcing the next race, and instead of the kids' names, I saw mine.

"Hey!" I called out to my friend. "We're up next! It's not the kids' race, it's the adults'!"

"Great!" he exclaimed, and he started putting on his own gear.

"Wait a minute!" I shouted. "I didn't watch the safety video yet! I don't know what I'm doing! What are the rules of the road? How do I drive this crazy . . ." My cries were drowned out as he threw a helmet on my head and strapped on my neck brace.

"Let's go!" he yelled, pointing to my car.

I climbed shakily in, figured out how to buckle my seatbelt, asked for help to adjust my gas and brake pedals, and attempted to look behind me, where my friend was revving his engine. I forced my face shield up, breathed in the exhaust fumes, and yelled back to him, "Seriously! What do I need to know? What is this like? *How will I know what to do?*"

Out of the corner of my eye, I could just make out my friend's grin through his helmet as he yelled back, *"Just don't come in last!"*

And the race began.

Leadership is about trust. But it's about trusting yourself—your instincts—and trusting the process as much as any other aspect of trust. You won't always know the rules of the road. Others will have more experience than you. Maybe someone you know well will encourage you to step into the unknown before you think you're ready—or maybe some unknown element will force you in that direction. You won't always have time to prepare. But you can still take a deep breath, rely on your instinct, and lean in to the curves up ahead.

No one else can lead the way you lead.

No one else has the purpose you do as a leader.

No one else can have the impact that you were designed to have on the world.

Embrace and celebrate who you are as an instinctual leader. Share the insights from the miles and find the joy in the moments that brought you to where you are today. Soak in the wisdom from the wildlife and wild places all around you. Lead in the way you were born to lead.

The animals and I are cheering you on.

Behind the Scenes with KiwiE

I couldn't believe I was in Alaska.

Pulling my car into the coastal city of Seward, my immediate goal was to get on the water in the most up-close-and-personal way possible. I passed by the large boats filled with tourists and instead chose a small boat that held no more than twelve guests plus the crew. The trip was created for wildlife photographers—specifically those searching for birds—and was perfect for me.

We all lugged our camera gear onto the boat along with extra clothing and overcoat layers to prepare for the day. The crew welcomed us and asked what we wanted to see, so they could do their best to accommodate our wishes. One by one, my fellow photographers named off various species of birds along with sea otters and a few other animals. Then it was my turn. "I'm up for anything, but I really want to see puffins!"

Silence fell, followed by hearty laughter from everyone onboard— except me.

I looked around, perplexed.

Finally, one crew member gathered his composure and said, stifling more laughter, "OK, ma'am. We'll do our best. But if we *don't* see puffins, there's definitely something wrong."

Feeling more than a little embarrassed, I gathered my stuff and chose a position at the front of the ship, camera poised and ready. We headed off onto the sea, turned the corner at the edge of the harbor, and . . .

Puffins!

Then more puffins!

Still more!

And up ahead, approaching the Northwest Glacier, you guessed it— even more puffins!

It was kind of like seeing squirrels outside your house, or maybe sparrows or robins, or pigeons, if you live in the city. Puffins really were everywhere, as everybody (but me) had known they would be.

I got my fill of puffins that day. And I was happy.

INSTINCTUAL LEADERSHIP FIELD GUIDE: FINAL REFLECTIONS

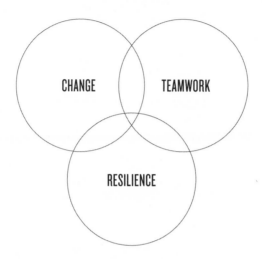

Now it's your turn! One last time, I invite you to reflect on the ideas, tools, case studies, and insights provided in this book, all centered on the leadership concepts of *change, teamwork,* and *resilience.* The following prompts will guide your reflections as you critically evaluate the relevant lessons and how they can best help you.

To target your answers and make them as immediately applicable as possible, I advise you to think about your answers only as related to your current role as a leader.

As you're reflecting, think of your own examples, case studies, and maybe even a few animal stories! Jot those down in the white spaces around these prompts. They will come in handy as you refer back to this book and pass along these lessons to others in your life.

This is how you truly personalize and implement the wisdom from the wild!

1. I believe I am an instinctual leader because:

   _____

   _____

   _____

   _____

2. My teammates/colleagues would say it is because:

   _____

   _____

   _____

   _____

3. My supervisor/manager/leader would say it is because:

   _____

   _____

   _____

   _____

4. My friends and family would say it is because:

   _____

   _____

   _____

   _____

5. My favorite teacher/mentor would say it is because:

_____

_____

_____

_____

6. The three words that best describe my individuality
   as a leader are:

   (1) _____

   (2) _____

   (3) _____

7. A time when I remember trusting my own gut is:

_____

_____

_____

_____

8. The two people who have most influenced my growth
   as an instinctual leader are:

   (1) _____

   (2) _____

9. If they were sitting in front of me, I would tell them:

_____

_____

_____

10. If I could accomplish one thing as a leader in the next twelve months, it would be:

_____

_____

_____

11. The unique way I will do this as an instinctual leader is:

_____

_____

_____

12. I would like my legacy as a leader to be:

_____

_____

_____

Now is the time—launch your instincts into action!

ACKNOWLEDGMENTS

To all the people who believed in me, took a chance, said yes when I put my hand up, and provided invaluable behind-the-scenes support and encouragement—you are the wind beneath my wings.

To all the people who challenged me, didn't give me the job, disagreed with my input, and didn't choose my ideas—your dissension caused me to take a second look, refine my approach, and learn the crucial life skill of how to agree to disagree.

To the leaders in every facet of life, whether corporate or nonprofit, government or community, who are driven by purpose, who face change day in and day out, who show up for their team, and who take a stand for resilience—thank you for your commitment to your people and your purpose. I am so proud of you.

To all the people working hard for the future of wildlife and wild places, whether you're on staff or volunteering, whether you're a scientist in the field or an accountant crunching numbers, whether your focus is global in scope or local and community-based—you are seen, recognized, needed, and appreciated.

To the team at Greenleaf Book Group—thank you for being as excited as I was about this project and by my side every step of the way. My amazing editors and proofreaders—Erin Brown, Amy McIlwaine, Susan Flurry, and Killian Piraro—your insightful questions and flow suggestions enhanced the reader's experience. Nette Pletcher, your attention to scientific detail ensured critical accuracy. Kevin Stone, your creative illustrations brought the animals (and plants) to life. Brian Phillips, you expertly captured this book's essence in the cover and design. Jen Glynn, your expert project management skills kept me laser-focused. Barry Banther, your vision of these ideas as a book (when they were still notecards on a table) connected me with Greenleaf. I am forever grateful to all of you for your unyielding belief in me and this message.

To Dr. Ann Haley MacKenzie and Dr. Don Kaufman, who first nurtured the possibilities of combining my love for nature, education, and leadership, and who helped forge pathways to make it possible—my projects from your college courses formed the basis of this book and still occupy a prominent place in my bookshelf today.

To Ted Beattie, a legend among leaders in the aquarium and zoo community, who challenged us to attain visionary goals and cultivated a welcoming and happy community for the people and the animals under his care—I was beyond fortunate to begin learning from you early in my career and it is the highest honor that you wrote the Foreword for this book.

To the person at the aquarium who forwarded me the phone call from the manager at Nabisco, inquiring about a possible team-building day—I will never forget standing at the front of the training room, kicking off the day of team-building activities with two of my mentors and watching my ideas first come to life.

To my parents, who encouraged me to go outside and lie in the grass to look at the clouds, who believe that all words—spoken or written—matter, and who taught me that leadership can take many shapes, but all are equally important—I understand now, as a mother, why you didn't let me take that skunk carcass our cat had hidden in the bushes into school for show-and-tell. But I still think it would have been awesome.

To my love, Michael, for all of the inspiration gleaned while camping, hiking, sailing, or any other outdoor adventure we found ourselves on—I can hear you cheering me on from Heaven and know you still walk alongside me every step of the way.

To my children, Tasman and Kepler, who stop to look at every egret with me, who come with me to the beach in hats and gloves when it's 40 degrees, and who indulge my need to explore under roots and in the mud no matter what I'm wearing—every sticky note, stuffed animal, mug of tea, glass of wine, cat for my lap, and piece of dark chocolate that you brought me was integral to the writing of this book.

And although I will continue to champion the work of aquariums, zoos, nature centers, wildlife rehabilitation centers, and conservation organizations of all shapes and sizes around the world, there are four that will always be my favorite: Cincinnati Zoo and Botanical Garden, Mote Marine Laboratory and Aquarium, Busch Gardens, and the John G. Shedd Aquarium. Your mark is indelible upon me, and a part of my heart will always be with you.

NOTES

Preface

1. Jared Diamond, "Playing God at the Zoo," *Discover Magazine*, March 1, 1995, https://www.discovermagazine.com/planet-earth/playing-god-at-the-zoo.

Author's Note

1. Rachel Treisman, "Barcelona Opera Reopens with an Audience of Plants," NPR, June 22, 2020, https://www.npr.org/sections/coronavirus-live-updates/2020/06/22/881943143/barcelona-opera-reopens-with-an-audience-of-plants.

Introduction

1. https://www.bostonglobe.com/metro/2016/04/14/new-england-aquarium-has-its-own-octopus-escape-story/3ShjEIp3tdIAqbLGPLtuSO/story.html.
2. https://ocean.si.edu/ocean-life/invertebrates/revealing-largest-octopus.

Part 1

1. National Soft Skills Association, "The Soft Skills Disconnect," February 13, 2015, https://www.nationalsoftskills.org/the-soft-skills-disconnect/.

Unbreakable Law #1

1. https://www.conservation.org/stories/11-facts-you-need-to-know-about-mangroves.
2. https://naturesacademy.org/florida-feature/florida-feature-white-mangroves-frankenstein-leaf/.
3. Jill Bailey, ed., *The Way Nature Works* (New York: Macmillan Publishing Company, 1992), 22.

Unbreakable Law #2

1. http://entnemdept.ufl.edu/creatures/urban/spiders/giant_crab_spider.htm.
2. https://www.smithsonianmag.com/smithsonian-institution/ask-smithsonian-how-do-spiders-make-webs-180957426/.

Unbreakable Law #3

1. https://www.worldwildlife.org/stories/how-long-do-sea-turtles-live-and-other-sea-turtle-facts.
2. https://www.nationalgeographic.com/news/2015/1/150115-loggerheads-sea-turtles-navigation-magnetic-field-science/.
3. https://oceanservice.noaa.gov/news/june15/sea-turtles.html.

Unbreakable Law #4

1. Michael Hehenberger, Zhi Xia, *Our Animal Connection: What Sapiens Can Learn from Other Species* (Singapore: Jenny Stanford Publishing Pte. Ltd., 2021).
2. https://www.theguardian.com/science/2013/jul/14/naked-mole-rat-cancer-research.
3. https://animals.sandiegozoo.org/animals/naked-mole-rat.
4. https://www.nature.com/articles/s41598-020-74.
5. https://www.scientificamerican.com/article/how-do-whales-and-dolphin/.929-6.

Unbreakable Law #5

1. Sindya N. Bhanoo, "On Savanna, Termites Are a Force for Good," *New York Times*, May 31, 2010, https://www.nytimes.com/2010/06/01/science/01obtermites.html.
2. Robert M. Pringle, Daniel F. Doak, Alison K. Brody, Rudy Joqué, "Spatial Pattern Enhances Ecosystem Functioning in an African Savanna," *PLOS*, May 25, 2010, https://journals.plos.org/plosbiology/article/info%3Adoi%2F10.1371%2Fjournal.pbio.1000377.
3. https://robertsrules.com/.

Unbreakable Law #6

1. https://ocean.si.edu/ocean-life/fish/tough-teeth-and-parrotfish-poop.
2. https://www.nationalgeographic.org/encyclopedia/coral/.
3. https://ocean.si.edu/ocean-life/invertebrates/corals-and-coral-reefs.
4. https://www.epa.gov/coral-reefs/basic-information-about-coral-reefs.
5. https://www.epa.gov/coral-reefs/basic-information-about-coral-reefs.

Part 3

1. https://www.nzgeo.com/stories/kea-the-feisty-parrot/.

Unbreakable Law #7

1. "All about Birds," The Cornell Lab, Cornell University, https://www.allaboutbirds.org/guide/Bald_Eagle/overview.
2. https://mote.org/news/article/remembering-the-shark-lady-the-life-and-legacy-of-dr.-eugenie-clark.

Unbreakable Law #8

1. https://zoologicalletters.biomedcentral.com/articles/10.1186/s40851-019-0133-3.
2. Stacey Dietsch and Elizabeth McNally, "Tomorrow's teams today: Build the capabilities needed to transform individuals, teams, and your organization," November 19, 2020, *McKinsey & Company*, podcast, 26:30, https://www.mckinsey.com/about-us/covid-response-center/leadership-mindsets/webinars/tomorrows-teams-today-build-the-capabilities-needed-to-transform-individuals-teams-and-your-organization.
3. Dietsch and McNally, "Tomorrow's teams today," November 19, 2020.
4. https://www.nationalgeographic.com/animals/invertebrates/group/sea-cucumbers/.

Unbreakable Law #9

1. https://cheetah.org/learn/about-cheetahs/.

INDEX

The Author in Her Natural Habitat

ABOUT THE AUTHOR

Julie C. Henry is an outdoor explorer and animal lover at heart who is continually learning and sharing leadership lessons inspired by wildlife and wild places.

A former zoo and aquarium senior leader, Julie is president of Finish Line Leadership, a training and consulting firm, and has worked with over fifty-five organizations across corporate, nonprofit, government, association, and community sectors. She holds Bachelor of Science degrees in both zoology and education (Miami University of Ohio), a Master of Arts in communication (University of South Florida), a Postgraduate Diploma in outdoor education (University of Otago, New Zealand), an Executive Program Certificate in sustainable business leadership (Green Mountain College), and has completed graduate coursework in winter ecology in Grand Teton National Park (Teton Science School). She was selected as a Fellow of the Toyota TogetherGreen program of the National Audubon Society, as part of the Disney's Animal Kingdom/World Wildlife Fund Biodiversity Leadership Institute, and as a Rotary Ambassadorial Scholar to New Zealand.

Julie has presented to over one million people across thirty-two states and six countries, in on-site and online settings ranging from auditoriums and ballrooms to boats, beaches, forests, theaters, boardrooms, and even underwater while feeding sharks and moray eels. She has yet to see a wolf in the wild, snorkel with whale sharks, or visit Antarctica, but continues to dream about those moments!

She lives in Sarasota, Florida, with her two children, whom she lovingly describes as her "zoo animal" and her "wild animal" due to each child's natural inclination toward life.